T0135548

UNIVERSITY OF FREIBURG

A New Robotic System for Visually Controlled Percutaneous Interventions under X-ray Fluoroscopy or CT-imaging

A DISSERTATION SUBMITTED TO THE
FACULTY OF APPLIED SCIENCES
OF THE ALBERT-LUDWIG-UNIVERSITY FREIBURG

for the degree
Dr. Ing.
Field of Microsystems Technology
By

MICHAEL H. LOSER

Freiburg im Breisgau, Germany
January 2005

Bibliographic information published by Die Deutsche Bibliothek

Die Deutsche Bibliothek lists this publication in the Deutsche Nationalbibliografie;
detailed bibliographic data is available in the Internet at http://dnb.ddb.de.

ISBN 3-8325-0845-7

Logos Verlag Berlin
Comeniushof, Gubener Str. 47,
10243 Berlin
Tel.: +49 030 42 85 10 90
Fax: +49 030 42 85 10 92
INTERNET: http://www.logos-verlag.de

Dean
Prof. Dr. Jan G. Korvink, Faculty of Applied Sciences

1. Reviewer
Prof. Dr. W. Menz, Department of Microsystem Technology

2. Reviewer
Prof. Dr. W. Burgard, Department of Computer Science

Chairman of Examination Board
Prof. Dr. H. Burkhardt, Department of Computer Science

Date of the Disputation
19 January, 2005

Ich erkläre hiermit, dass ich die vor-
liegende Arbeit ohne unzulässige Hilfe
Dritter und ohne Benutzung anderer als
der angegebenen Hilfsmittel angefertigt
habe. Die aus anderen Quellen direkt
oder indirekt übernommenen Daten
und Konzepte sind unter Angabe der
Quelle gekennzeichnet. Insbesondere
habe ich hierfür nicht die entgeltliche
Hilfe von Vermittlungs- oder Bera-
tungsdiensten (Promotionsberaterinnen
oder Promotionsberater oder anderer
Personen) in Anspruch genommen.
Niemand hat von mir unmittelbar oder
mittelbar geldwerte Leistungen für Ar-
beiten erhalten, die im Zusammenhang
mit dem Inhalt der vorgelegten Dissertation stehen. Die Arbeit wurde bisher
weder im In- noch im Ausland in
gleicher oder ähnlicher Form einer anderen Prüfungsbehörde vorgelegt.

Erlangen, den 25.01.2005

Michael H. Loser

Abstract

Advances in technology have resulted in substantial changes in medicine. Minimally invasive image-guided interventions have been developed, employing modern cross-sectional imaging techniques combined with novel high-tech therapeutic devices. This new surgical environment allows for sophisticated visualization, localization, access, and control of a surgical instrument, fully integrated and adapted to the needs of a human operator. Modern operating rooms increasingly employ real-time imaging systems to guide during interventions. Typically, *X-ray fluoroscopy* or *CT-imaging* is applied, since both modalities provide real-time image feedback with appropriate quality and convenient access to the patient during imaging. However, although these modalities are appealing and appropriate for image-guided percutaneous procedures, both of them use ionizing radiation. And, since image-guided interventions tend to become longer and more complicated, the *radiation dose* especially for the operator, is becoming an increasing source of concern.

An answer to this problem is presented in this actual thesis: an *interventional multimodal workplace with a micro-robot* for image-guided percutaneous procedures. This thesis presents the development of a *miniaturized robotic system* to support in needle placement procedures under X-ray fluoroscopy or CT-imaging. Due to its very compact and light design, the micro-robot is optimized for usage together with an X-ray C-arm or inside a CT-scanner, where a bulky robot is inappropriate, especially together with a stout patient or long, stiff instruments like a biopsy needle.

In addition, a novel *automatic image-guided control algorithm* based on "visual servoing" has been developed for the robot. This means, that the robot is directly controlled by real-time imaging feedback provided by the X-ray C-arm or the CT-scanner. Visual servoing is well established in the field of industrial robotics, when using CCD cameras. This approach has been adapted and optimized to guide a puncture needle using X-ray fluoroscopy or CT-imaging. This resulted in a simple and accurate method for positioning a needle with respect to a deeply seated target inside a patient. In contrast to all other existing methods - referred in literature - for robot-supported percutaneous procedures using X-ray or CT-imaging, this approach requires no prior calibration or registration. Therefore, no additional sensors (infrared, laser, ultrasound, etc), no stereotactic frame and no additional calibration phantom is needed, as in other methods.

In this thesis, the *precise placement of a puncture needle* for percutaneous procedures is the medical *key application* which is implemented in the robot-supported imaging workplace. After development and implementation of the image-based robot control, the efficacy and precision of this new approach is demonstrated by several phantom and cadaver studies. The first evaluation of this control approach using X-ray-fluoroscopy imaging has been performed by the author with three different medical robots at the Computer Integrated Surgery (CIS) Lab at the Johns Hopkins University, Baltimore, USA. Further experiments were conducted with X-ray fluoroscopy and CT-imaging at the Siemens Medical Solutions laboratories, Germany. These experiments were conducted with the newly developed micro-robot, which is presented in this thesis. A number of cadaver trials were performed in order to prove the new micro-robot and the image-based control under more realistic clinical conditions. In these test series a needle placement accuracy of about ±1.5mm has been achieved with X-ray fluoroscopy. Using CT-imaging, the accuracy of the needle placement procedure was about ±1.6mm.

This thesis describes the first use of visual servoing for automatic X-ray- and CT-guided needle placement procedures. This technique allows for automatic needle alignment with a deeply seated target inside a patient, without prior calibration or registration of the robot or the imaging system. The promising results present the novel visual control approach as an appropriate alternative to other needle placement techniques requiring cumbersome and time consuming calibration procedures. In addition to increased accuracy, the system accelerates the puncture procedure and reduces the X-ray exposure for both patient and surgeon.

Zusammenfassung

Technologischer Fortschritt führte in den letzten Jahren zu grundlegenden Veränderungen in der Medizin. Es wurden z. B. minimalinvasive bildgestützte Verfahren entwickelt, die radiologische bildgebende Systeme mit neuartigen high-tech Instrumenten kombinieren. Oftmals wird hierfür röntgenbasierte Bildgebung eingesetzt, wie z. B. konventionelle Röngendurchleuchtung oder die Computertomographie. Da aber bildgestützte Eingriffe zunehmend komplizierter und langwieriger werden, gibt die resultierende Strahlenbelastung - insbesondere für den Operateur - oft Grund zur Besorgnis.

Eine Antwort auf dieses Problem ist Kern der vorliegenden Dissertation: ein *interventioneller, multimodaler radiologischer Arbeitsplatz ausgestattet mit einem Mikroroboter für bildgestützte perkutane Interventionen*. Diese Arbeit präsentiert die Entwicklung eines Mikroroboters, der für Nadelpunktionen unter Röntgendurchleuchtung oder CT-Bildgebung eingesetzt wird. Wegen seines sehr kompakten Aufbaus ist der Nadelführungs-Roboter bestens geeignet für Punktionen im Zusammenspiel mit Röntgen C-Bögen oder in CT-Scannern.

Darüber hinaus wurde im Rahmen dieser Arbeit ein neuartiger, bildgestützter Steuerungsansatz für den Roboter entwickelt (visual servoing), mit dessen Hilfe die Punktionsnadel automatisch auf eine zuvor vom Arzt definierte Zielstruktur ausgerichtet werden kann. Das heißt, der Roboter verwendet direkt die Bilddaten des bildgebenden Systems für seine Bewegungssteuerung. Dieser Ansatz findet bereits im Bereich der industriellen Robotik Verwendung und wurde im Rahmen der vorliegenden Arbeit auf den medizinischen Bereich übertragen und speziell für Nadelpunktionen optimiert. Im Gegensatz zu anderen bisher bekannten Methoden, die in der Literatur beschrieben werden für Roboter unterstützte perkutane Interventionen unter Röngen- oder CT-Bildgebung, benötigt der hier präsentierte Ansatz keine vorherige Kalibrierung oder Registrierung des Roboters oder des bildgebenden Systems. Folglich werden keine zusätzlichen Sensoren (Infrarot, Laser, Ultraschall, o. ä.), keine stereotaktischen Rahmen oder Kalibrierphantome benötigt, wie dies mit anderen Methoden der Fall ist.

Die erste Erprobung der bildgestützten Steuerung erfolgte mit drei verschiedenen Robotern am Computer Integrated Surgery (CIS) Lab an der Johns Hopkins University, Baltimore, USA. Weitere Experimente mit Röntgen- und CT-Bildgebung erfolgten in Laboratorien der Siemens Medical Solutions in Deutschland, wobei hierfür der eigens entwickelte Mikroroboter zum Einsatz kam. In zahlreichen Kadaverversuchen wurde eine Positioniergenauigkeit der Nadel von etwa ±1.5mm unter Röntgenbildgebung und etwa ±1.6mm unter CT-Bildgebung erreicht.

Diese Arbeit beschreibt die erstmalige Anwendung einer bildgestützten Steuerung (visual servoing) für automatische Nadelpunktionen unter Röntgen- oder CT-Bildgebung. Dieses Verfahren kommt ohne vorhergehende Kalibrierung oder Registrierung des Roboters zum Bildgebenden Systems aus. Die vielversprechenden Ergebnisse stellen den neuartigen bildgestützten Steuerungsansatz als eine brauchbare Alternative dar, verglichen zu anderen Nadelführungstechniken, die mühsame und zeitaufwendige Registrierungsprozeduren erfordern. Neben erhöhter Genauigkeit und verkürzter Dauer der Intervention, reduziert der Roboter zusätzlich die Strahlenbelastung sowohl für den Patienten als auch den Chirurgen.

Acknowledgements

First and foremost, I would like to thank my advisor, Prof. Dr. Wolfgang Menz, for his unbroken support during the seven years of my Ph.D. Moreover, I want to thank my other dissertation committee members, Prof. Dr. Hans Burkhardt, Prof. Dr. Peter Woias and Prof. Dr. Wolfram Burgard for their kind disposition to review my thesis and for the great discussions during my disputation.

Thanks to Dr. Sigfried Bocionek and Dr. Rainer Graumann who gave me the opportunity to work at the Siemens Medical Solutions R&D Department and enjoy an exceptional research environment and facilities. Furthermore, I want to thank all my former colleagues at the R&D Department who gratefully supported me during my work. In particular special appreciation to Dr. Oliver Schütz, Dr. Helmut Barfuß, and Norbert Rahn. On a personal level, I want to thank my Ph.D. colleagues at the R&D Department, especially Joachim Tork, Matthias Mitschke and all the other folks which let us have a great time.

Special thanks to Prof. Dr. Russell H. Taylor, who gave me the great opportunity to work in his group for half a year at the Computer Integrated Surgery Lab at the Johns Hopkins University, USA. Thanks to all the members of the CIS Lab, especially Andy Bzostek, Rajesh Kumar and Jianhua Yao, who provided me great support in building up the robot setup and helped me to conduct my first experiments in visual servoing based needle placement. Thanks to Prof. Dr. James Anderson and Aaron Barnes who advised and supported me during the X-ray experiments at the Johns Hopkins Medical School. Furthermore, I want to thank Sean Hundtofte, Randy Goldberg and all the other folks from the CIS Lab, who made my time at the Johns Hopkins University not only scientific but also greatly enjoyable.

Great thanks to Dr. Alok Gupta at the Siemens Corporate Research, Princeton, USA, who supported me during the collaboration with the Johns Hopkins University and my work at the CIS Lab.

Last but not least, special thanks to Prof. Dr. Nassir Navab who contributed the original ideas and the theoretical framework on fluoro-servoing needle placement. Thank you Nassir, not only for your great assistance I have received throughout my work, but also for treating me as a colleague and friend.

Contents

Abbreviations and Nomenclature

The abbreviations, special terms and most commonly used symbols used in this thesis are listed in this section.

Abbreviations:

A/D	analogue digital converter
BRW	Brown-Robert-Wells stereotactic system
CAN	controller area network (serial bus standard)
CAS	computer assisted surgery
CCD	charge-coupled device (an integrated circuit for digital photography)
CIS	computer integrated surgery
CT	computed tomography
D/A	digital analogue converter
DC	direct current
DLL	dynamic link library
dof	degree-of-freedom
DSP	digital signal processor
e.g.	for example (lat.: exempli gratia)
FDA	Food and Drug Administration
i.e.	that is (lat.: id est)
I/O	input/output, refers to the interfaces of a system
IR	infra red
JHU	Johns Hopkins University
LARS	Laparoscopic Assistant Robot System
LED	light emitting diode
MIP	maximum intensity projection
MPR	multi planar reformation
MRC	modular robot control (JHU software library for robot control)
MRI	magnetic resonance imaging
MS	Microsoft™
NTSC	National Television Standards Committee (video standard)
OR	operating room
PAKY	Percutaneous Access to the KidneY
PAL	Phase Alternating Line (video standard)
PID	proportional, integral, derivative (controller)
RAM	random access memory (type of computer memory)
RCM	remote center of motion (type of robot developed by JHU)
RGB	red, green, blue color model
TTL	transistor-transistor logic
TV	television

special terms:

CARE Vision	CT fluoroscopy scan mode of the Siemens Somatom CT scanners
C-arm	here: short form of X-ray C-arm system
CT fluoroscopy	CT scan mode providing nearly 'real-time' CT imaging
gantry	the mechanical body of a CT scanner
intervention	small surgery
pose	position and orientation of an object, expressed e.g. as matrix
sequence-CT	CT scan mode where only a single slice of the patient is acquired. The patient is not moved during image acquisition.
spiral-CT	CT scan mode where the the patient on the CT table is moved through the CT gantry during imaging (acquisition of a 3D volume)
table top	the CT table top ist the moving part of the CT table on wich the patient is positioned and horizontally moved into the CT gantry
visual servoing	image-based control, e.g. of a robot
X-ray C-ram	an X-ray fluoroscope in the shape of a 'C', with the X-ray generator and the imaging detector on opposite ends of the 'C'.
X-ray fluoroscope	X-ray based medical imaging system (typically mobile) which shows the permanently acquired X-ray images in real-time on a monitor

most commonly used symbols:

(some symbols are overloaded in which case their context must be used to disambiguate them)

δ	deflection / deviation angle
Δ	deviation
γ	angle of needle in image
λ	cross-ratio
η	relation between millimeter and pixel size in image
τ	threshold for image feature extraction
π, Π	plane
α, β	angles of motor axes
ψ, θ, ϕ	Euler angles
d	CT slice thickness / spring constant
D	diameter
E	skin entry point of needle insertion / elastic modulus
F	force
f	image feature
I	geometrical moment of inertia
L	length / needle length
M	bending moment, marker location
N	needle coordinate frame / needle pose
n	needle vector (along needle axis)
n	normal vector
p	2D pixel location in the image
$q(x)$	load distribution
R	rotation matrix
S	optical center / center of gravity
T	target location / translation vector
u, v	image coordinates
$w(x)$	bending of a cannula (elastic line)
${}_{B}^{A}[\ ]$	transformation matrix converting pose A into pose B

Chapter 1
Motivation

1.1 Introduction

Advances in several technologies within the last decades have resulted in substantial changes in medicine, especially in the field of radiology and surgery. Much of this progress has been driven by improvements in computer technology and the widespread adoption of digital techniques for data acquisition, processing and display. Doubtless, milestones have been the introduction of CT and MRI scanners into clinical routine in the 1970s and 80s. It provided digital high-resolution, cross-sectional images with real 3D information for the first time. This novel capability of sophisticated *3D imaging* had not only great impact on the quality of diagnostics but also enabled precise planning of surgical interventions. Furthermore, the digital image data allowed for digital image processing, 3D reconstruction, simulation and modeling, and even computer aided diagnosis [175].

In the last two decades medical imaging increasingly became an integral part of the surgical process. 3D imaging is employed for planning of the surgical intervention, for positioning or guiding a therapeutic tool during the surgery, and verifying the therapeutic outcome [67].

In *conventional surgical procedures* the surgeon basically relies on the *direct visibility* of the patient's anatomy. In the operating room the surgeon depends on what he or she is seeing, mostly supported by radiographic films that are fixed to the light-box in the OR (compare figure 1.1-A). These radiographic films are showing the interesting anatomical site of the patient and have been acquired by the radiology department before surgery (*preoperatively acquired images*). During the intervention the surgeon cannot see beyond the exposed surface. Within the constraint of the surgical opening, the exposed visible field is mostly insufficient to comprehend the entire anatomy. Limitations of this *open surgical exploration* have several consequences. Since preoperative radiographs provide only a rudimentary anatomical overview, localization is not geometrically accurate, especially not in 3D. Therefore, the definition and execution of exact trajectories for targeting is impossible.

The integration of *computer assistance* during the surgery on basis of digital, preoperative 3D image data is leading to a new specialty in surgery called *computer assisted surgery* (CAS, compare figure 1.1-B). CAS is creating a link between the preoperative data acquisition (e.g. 3D models from CT or MRI scans) and the execution of the surgery [114]. Tools that are required to provide this link are *stereotactic-* or *navigation-systems* which allow intraoperative guidance of e.g. a surgical tool. Well-known are for example optical navigation systems on base of an infrared (IR) stereo camera system, as shown in figure 1.1-B. CAS intends to help the surgeon improving the accuracy and safety of the surgical procedure.

Although, CAS provides improved accuracy with which a surgical procedure can be performed, it has a significant drawback: CAS relies on preoperatively acquired image data for guiding a surgical tool, and assumes that there are no changes in the anatomical structures after image acquisition and during the surgery. Otherwise the preoperative image data of the patient would not be identical to the anatomy of the patient in the OR. However, almost all surgical procedures entail movement of tissue, through the use of retractors, by surgical resection or due to the drainage of fluids. One exception might be the surgery on rigid structures

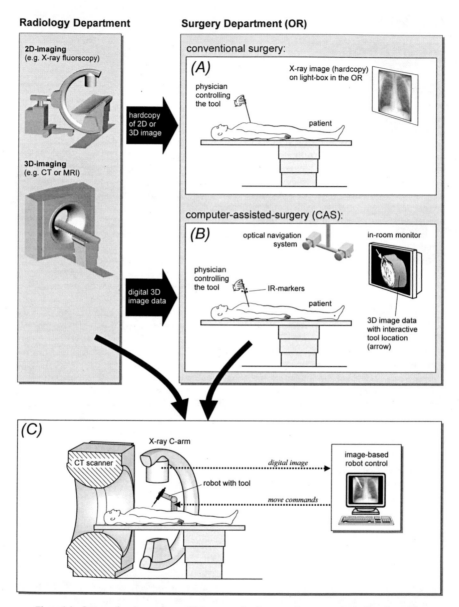

Figure 1.1: Interventional procedures: (A) In conventional surgery the operator basically relies on the *direct visibility* of the patient's anatomy, only supported by preoperatively acquired radiographs. (B) Computer assisted surgery (CAS) allows intraoperative guidance, e.g. of a surgical tool, based on preoperative 3D image data. This is provided by stereotactic or navigational tools. (C) Employing real-time intraoperative imaging for the automatic control of a surgical robot leads to the novel scenario presented in this thesis.

as bones e.g. in orthopedic interventions. But it is obvious that especially for soft tissue procedures, the assumption that there is no tissue movement between imaging and the surgical act is questionable [78].

This drawback of CAS is leading to the need for *real-time imaging* (*intraoperative imaging*) during the surgical procedure. These real-time images provides updates about the patient's anatomy or the changing position of movable organs. Furthermore, it represents the position of instruments relative to the target structure, and thus establishes directly the required relationship between the instrument and the patients anatomy in the image, even *without prior registration* (compare page 9). For example, multiple projections of an X-ray fluoroscope [110][1] or the acquisition of multiple CT slices [2] can provide a comprehensive description of 3D relationships between the instrument and the target. This type of interactive visualization is sufficient for guidance of most percutaneous biopsy and intravascular interventional procedures [78].

Unfortunately, the conventional operating room usually does not provide quality images as can be obtained by radiology sites [18]. On the other hand, a conventional radiology site does not meet sterility guidelines to perform surgical procedures. However, there are many minimally invasive interventions, that are already performed by radiologists under direct image guidance (*interventional radiology*). Examples are intravascular procedures and percutaneous needle placement procedures as shown in principle in figure 1.2.

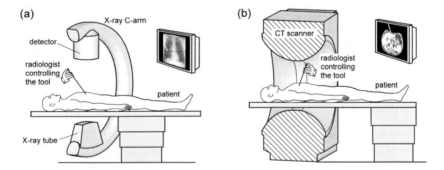

Figure 1.2:   Interventional radiology: Real-time imaging control is provided (a) with an X-ray C-arm, (b) with a CT-scanner. In both cases in-room monitors are showing the image of the patient's anatomy together with the surgical tool (here a needle). These images allow the interventional radiologist to properly manipulate the tool inside the patient without direct visibility of the surgical area.

There are currently various concepts discussed for employing sophisticated imaging modalities during surgery, including ultrasound, CT, or MRI. Radiologists, for example, are in favor of the advanced interventional radiology site appropriate for surgical interventions [164], while surgeons prefer OR sites with integrated imaging modalities like angiography systems or CT scanners [100]. Melzer, *et al.* [104], sees the future in close interdisciplinary collaboration between these two specialties with the establishment of image-guided surgical interventions.

1.2 The Novel Approach Presented in this Thesis

A combination of these two approaches, (i) directly image-guided procedures performed by interventional radiologists, and (ii) sophisticated techniques applied by surgeons in CAS procedures, *leads to the basic topic covered in this thesis*. The proposed scenario results in an *interventional multimodal workplace with a micro-robot*. This novel concept presented in this thesis is shown in principle in figure 1.1-C. The radiological workplace shall allow the physician to perform surgical procedures with real-time imaging control provided by an X-ray C-arm or a CT scanner. Both of these two imaging modalities are well established in the medical field. And since they provide convenient access to the patient during imaging and show adequate image quality and real-time imaging capabilities, both modalities are appropriate for *directly image-guided interventions*. In order to meet sterility demands, it is required to cover the X-ray C-arm and the CT scanner with surgical drape during the intervention.

The combined use of these two imaging modalities in one workplace has already been realized at some radiology sites [53][74]. These installations are so-called *hybrid systems*. They combine the advantages of *cross-sectional* (CT) with *projective imaging* (C-arm) and allow for improved control during e.g. percutaneous drainage or intravascular interventions [89] [134][13][90].

However, a basic disadvantage associated with X-ray fluoroscopy and CT imaging is the *radiation exposure* for the operator, especially if the intervention is performed under direct imaging. For this reason, a *robotic manipulator* is proposed for remotely controlled or even automatic guidance of a surgical tool. The physician could stand away from the primary X-ray beam or may even leave the operating room during the intervention.

A special *image based control* (visual servoing) has been developed and implemented for the robot, which allows semi-automatic and uncalibrated alignment of a surgical tool with a target under real-time image guidance: real-time images are acquired, analyzed and directly taken to achieve a surgical task with the manipulator, for example the alignment of a puncture needle. A distinct advantage of this real-time visual control is - over common CAS procedures - that *no prior calibration or registration* of the robot or the imaging modality is required. This means that no additional stereotactic or navigation systems are needed.

An uncomplex but challenging minimally invasive task is the precise placement of a puncture needle, e.g. for percutaneous biopsy, drainage, or tumor ablation. Needle placement is commonly performed manually and, especially in high risk areas, under direct image guidance. Therefore, *needle placement* is particularly suitable for the implementation into the novel imaging workplace and is identified as the *key application to be investigated in this thesis*. After development and implementation of the novel, image based control for the robot the efficacy and precision of this new approach will be demonstrated by several phantom and cadaver studies. This thesis describes the first use of visual servoing for automatic X-ray- and CT-guided needle placement procedures.

1.3 The Structure of this Dissertation

Figure 1.3 shows the procedure followed in this thesis. After a small market analysis, which has been conducted by the author, the *user needs* for an image guided robotic system have been specified. Furthermore, after consulting the interviewed physicians 'needle placement' has been defined as the key-application for the presented image based control approach, which is implemented and validated in this thesis. For this application of interest, the desired improvements and possible new surgical procedures have been formulated in a *clinical specification*. From this clinical specification, a *technical specification* has been derived in which

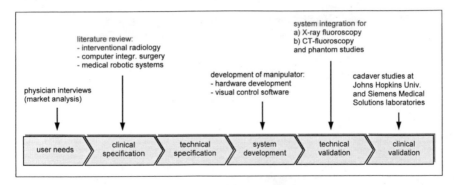

Figure 1.3:   The approach followed in the presented project for the development and validation of a robotic manipulator for automatic image-guided needle placement.

each of the required functions of the system were specified and translated into hardware and software requirements to be realized for the robotic system. After development and assembly of the manipulator and programming of the image based control software, the setup with integrated imaging modalities has been tested for a *technical validation* under predefined conditions in a phantom study. Based on the results of these first experiments the system has been continuously improved. Ensuing cadaver studies at the Johns Hopkins University and the laboratories at Siemens Medical Solutions, Germany, demonstrated the system capabilities under more clinical conditions within the scope of a *clinical validation*.

Over the last two decades many different techniques have been developed to support image-guided needle placement procedures. The approaches range from simple manual techniques up to the employment of robotic systems. Most of these systems apply a variety of different image guidance and registration approaches.

In order to appreciate the value and benefit of the novel approach presented in this thesis, it is essential to understand the medical background in this field. For this reason chapter 2 gives an overview of different approaches in computer assisted surgery (CAS) as well as several needle placement techniques commonly used in interventional radiology employing X-ray fluoroscopy or CT imaging. After a brief introduction to the model of a general purpose robotic manipulator, chapter 3 gives a survey of different types of medical robots applied to neurosurgery, orthopedics, and minimally invasive surgery.

After the introductory chapters 2 and 3, the author presents his contributions to this thesis. For a better understanding of the user needs in image-guided procedures, chapter 4 presents the results of a market analysis performed by the author. In several *physician interviews* the potential key-applications for the proposed automatic image-based approach could be identified in the field of radiology and orthopedics.

Chapter 5 presents the *development of the novel needle guiding robot* which is appropriate for usage inside a CT-scanner or in conjunction with an X-ray C-arm. All components of the robot are described, followed by a validation of the mechanical stiffness and manipulator accuracy. Furthermore, the kinematic model of the robotic manipulator is formulated.

Essential in this new approach is the use of a *visual control technique* (visual servoing) to manipulate the needle. This basic topic of the thesis is presented in chapter 6. The two approaches for automatic image-guided needle alignment have been developed for X-ray fluoroscopy and for CT-imaging. The presented image-based control algorithms allow alignment of a surgical tool with a target without prior calibration or registration of the robot or the imaging system. Finally, the development and implementation of all hardware and software components of the *visual servoing workstation* is described.

Chapter 7 introduces the experimental setup and presents the needle placement experiments under *X-ray imaging* performed at the Johns Hopkins University and the Siemens Medical Solutions laboratories. Although, the testing took place under laboratory conditions and not in clinical surrounding, the use of pig organs as test object for needle placement allowed to achieve quite realistic results.

The experimental setup with a CT scanner and all tests in automatic needle placement using *CT-imaging* are presented in chapter 8. Again, the final needle puncture experiments have been performed on pig organs.

All experimental results end experiences with the novel needle guiding robot are discussed in chapter 9. Several issues are studied, including precision, safety, radiation exposure, and costs. Furthermore, suggestions for further work are given. Since this project has generated much follow-up work in other research groups, e.g. at the Johns Hopkins University or the Georgetown University Hospital, the last section of chapter 9 deals with subsequent related work of other research groups.

Chapter 10 closes this thesis with a summary and concluding comments. Literature referred in this thesis is summarized in a reference list in chapter 11. The appendices contain some additional supporting material.

Chapter 2
Medical Background

In this thesis, the *precise placement of a puncture needle* is the *key application* which is implemented in the robot-supported imaging workplace, proposed above in section 1.2. Needle placement is commonly performed manually and, especially in high risk areas[1], with the help of real-time imaging.

Over the last two decades many different techniques have been developed to support image-guided needle placement procedures. The approaches range from simple manual techniques up to the employment of robotic systems. Most of these systems apply a variety of different image guidance and registration approaches. In order to appreciate the value and benefit of the novel approach presented in this thesis, compared to existing methods, it is essential to understand the medical background in this field.

Therefore, this chapter gives an overview of approaches in computer assisted surgery (CAS) as well as different needle placement techniques commonly used in interventional radiology employing X-ray fluoroscopy or CT imaging.

2.1 CAS - Computer Assisted Surgery

In the last two decades medical imaging increasingly became an integral part of the surgical process. Imaging is not only used for diagnostics but is also employed for planning of the surgical intervention, for positioning of the therapeutic tool in the preferred location, for guiding the intervention, and verifying the therapeutic outcome [67]. However, the surgeon has to deal with the basic problem of how to apply imaging during surgery.

In conventional surgical procedures the surgeon basically relies on *direct visibility* of the patient's anatomy. Within the constraint of the surgical opening, the exposed visible field is mostly insufficient to comprehend the entire anatomy. Furthermore, precise localization in 3D or the execution of exact trajectories for targeting is impossible.

Neurosurgery has been the first medical discipline dealing extensively with this problem. Already in 1908, Horley and Clark described a special apparatus consisting of a mechanical frame system attachable to a patients head [69]. This device, the first *stereotactic head-frame*, incorporated a Cartesian coordinate system for movement of a probe, which could be introduced into the brain. Since that time many frame-based stereotactic systems have been developed. While basically conventional X-ray radiography was used to relate the stereotactic frame position to the anatomy of the patient, the integration of CT imaging in the late 1970s, provided a much more powerful tool for neurosurgeons [180]. Several groups have developed CT-compatible stereotactic head frames for intracranial operations (compare figure 2.1). Although, the brain is ideally suited for such procedures because of the availability of rigid skull fixation and the absence of physiologic motion, the use of these frames presents the surgeon with certain problems: the manual adjustments required by the frame are time-consuming and error prone when frequent trajectory manipulations are required. Furthermore, a frame between surgeon and patient can be an awkward arrangement.

However, the introduction and dissemination of new digital technologies in the ensuing decades and their utilization in the medical field had strong impact on traditional frame-based

---

[1] High risk areas are e.g. regions close to large vessels or nerves.

stereotaxy. Various improvements lead to novel interactive image-guided stereotactic systems, employed in a novel medical field called *computer assisted surgery* (CAS).

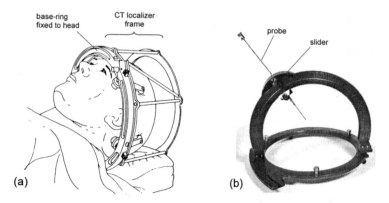

Figure 2.1: The Brown-Roberts-Wells (BRW) stereotactic system for CT. (a) The base-ring is attached to the patients head and defines the stereotactic coordinate system. During CT scanning the localizer frame is mounted on the base-ring. It consists of different parallel and diagonal carbon rods, which get intersected by the CT-scan plane and show up in the image as small dots around the patients head. The location of these points in the image provide sufficient mathematical data to map every pixel from the CT slice into the 3D coordinate system of the stereotactic base-ring.[2] (b) Later in the operating room the arc system is attached to the base-frame and is used to adjust the computed probe trajectory by moving the slider and the arc.[3]

The idea of CAS is to create a link between the preoperative data acquisition (e.g. 3D models out of CT or MRI scans) and the execution of the surgery. CAS implies methods that allow intraoperative navigation in the surgical field based on preoperative images in conjunction with *frame-based* or *frameless stereotactic systems*. The goal is to improve the accuracy with which a given surgical procedure can be performed compared to conventional methods. Furthermore, the performance of certain surgical interventions, especially in high risk areas, got only possible with the support of computer guidance or robotic tools. Since the beginning of computer assisted interventions in the 1980s [149], CAS has been applied to several medical fields like neurosurgery, orthopedics, minimally invasive surgery, craniofacial surgery, and others. Although, the needs and conditions of applications in these medical fields may vary a lot, the principle of how CAS is employed in the interventional workflow, is basically the same. The workflow in CAS is mostly categorized in: *(a) surgical planning and simulation, (b) registration,* and *(c) intraoperative tracking and visualization.*

Surgical planning and simulation

After preoperative acquisition of cross-sectional CT or MRI scans and subsequent *3D image reconstruction*, the obtained 3D volume of the patient's anatomy (*3D patient model*) is used for planning of the surgical intervention (compare figure 2.2). In case of tumor surgery in the brain for example, the goal is to define the safest possible approach with the least possible damage to normal tissue [78]. Taking real physical tissue properties into account, soft tissue

---

[2] Reprinted from: *Tumor Stereotaxis*, P. J. Kelly, W. B. Saunders Company, Philadelphia, 1991.
[3] Reprinted from: *Advanced Neurosurgical Navigation*, E. Alexander, et al. (Ed.), Thieme, New York, 1999.

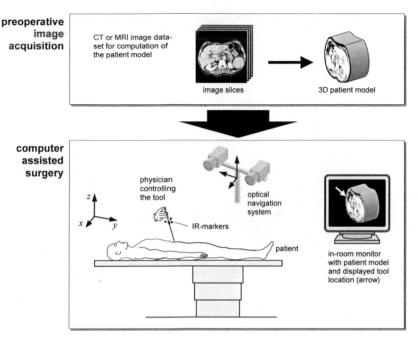

Figure 2.2: Workflow in CAS procedures. After preoperative CT or MRI image acquisition and ensuing 3D image reconstruction a virtual 3D patient model is generated. Provided with real physical tissue properties, this model can be used for simulation of the surgical procedure. After registration of the 3D patient model to the real patient in the OR, the patient model can be used together with a tracking system for interactive instrument guidance.

behavior of the reconstructed 3D patient model can be *simulated*. This deformable patient model allows appropriate simulation of cutting, retraction, drainage or resection procedures even before the real surgery [18][78].

Registration (image-to-patient registration)
The preoperatively acquired and processed 3D patient data can subsequently be employed in the operating room for *intraoperative guidance* of surgical tools. This requires matching of the 3D patient model with the corresponding anatomy of the patient positioned on the operation table (physical space). Matching of these two coordinate frames is called *image-to-patient registration* or just *registration*, and is usually accomplished by correlation of corresponding 3D-features, like characteristic points or shapes, on both the reconstructed image-model and the real patient [94][78]. A comprehensive overview of different registration methods in medical stereotaxis is given by Maciunas [99] and Lavallée [93].

The simplest registration method is the so-called *paired-point matching* [18], which means the correlation of point markers (fiducials) on the patient that can be seen both in the image-data as well as on the real patient. There are various methods existing for determination of these corresponding points. In the original *frame-based stereotactic method* a stereotactic frame is rigidly fixed to the patient during preoperative scanning and surgery (see figure 2.1). After scanning the patient's head together with its stereotactic frame, the 3D patient model can be described in the frame coordinate system e.g. by using a 'localizer frame' (compare figure 2.1-a). During surgery the same frame is equipped with movable arcs holding and positioning the instruments in defined orientations with regard to the stereotactic frame. Thus, the stereotactic frame represents a static link between the image and the instrument coordinate system (figure 2.1-b) [99][78][37].

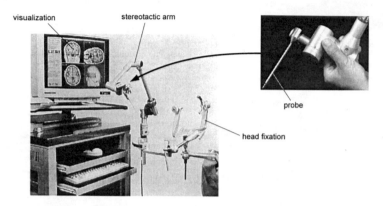

Figure 2.3:   Frameless stereotaxis with an arm-based system (Neuronavigator [169]). The arm has six degrees-of-freedom and is attached to a Mayfield head fixation device and connected to a computer workstation positioned beside the operating room table.[4]

However, frame-based systems proved to be cumbersome and have been almost entirely replaced by *frameless stereotactic methods* [19][163]. Early *arm-based systems* for example are holding the surgical instrument while tracking its tip via precise joint encoders [168][59]. Figure 2.3 shows an example of a stereotactic arm system. The basic function of this arm is to obtain the tip location within the surgical field and to translate it into the image coordinate frame of the 3D patient model displayed on a monitor. This requires registration of the image dataset to the real patient, which is achieved by touching at least three fiducial points or anatomical landmarks on the patient's head with the probe tip. Subsequently, the *corresponding points* have to be identified in the image dataset on the monitor with the mouse. Now, the *transformation matrix* between the image coordinate frame and the patient's head is automatically computed and can be employed to display the surgical tool in the image dataset on the monitor during surgery [169].

An alternative to the arm-based systems are *optical tracking systems* (stereo camera systems). They are tracking the instrument by detecting LED-markers that are fixed to it (see figure 2.4). This allows to precisely compute the location of the instrument tip in the coordinate frame of the optical tracking system. In case of patient registration with frameless stereotactic systems a pointer is tracked for determination of the fiducial markers or anatomic landmarks on the patient [78]. The registration procedure is identical to that described above.

Further frameless tracking devices for intraoperative guidance are *electromagnetic sensor systems* [71][150]. These systems allow to detect electromagnetic micro-coils (marker) through the body of the patient and therefore need no direct visible contact between sensor and marker as in optical tracking systems. Electromagnetic position tracking systems are predestinated for guiding flexible catheters or other instruments inside the patient's body.

---

[4] Reprinted from: *Advanced Neurosurgical Navigation*, E. Alexander, et al. (Ed.), Thieme, New York, 1999.

There exist several other types of sensors for acquisition of intraoperative registration data. Research groups have reported on the use of X-ray images [92], ultrasonic sensors [14], optical range imagers [62][147], laser scanners [62], video cameras [58][15], and even robots [87].

The use of a *robot* for intraoperative registration is appealing, because after the registration it can be directly used to perform certain surgical tasks. After a pointer at the robot's end-effector would be guided to the set of fiducials on the patient for *robot-to-patient registration* (basically the same procedure as described above) the robot could subsequently manipulate a surgical tool according to the preoperative plan.

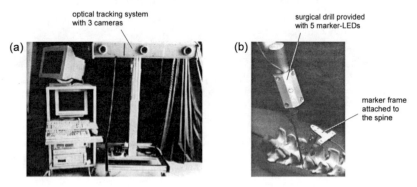

Figure 2.4: Frameless stereotaxis with an optical tracking system (OPTOTRACK®). [5]

Intraoperative guidance

Normally, the same tracking tools used for the registration procedure are employed for intra-operative guidance as well. LED-markers are mounted on the surgical instruments for tracking the path during the operation (see figure 2.4-b). After successful registration of the patient to the preoperatively acquired image dataset, the 3D-model is accurately displayed together with additional geometrical information, like the planned access trajectory and the current position of the surgical instrument, on monitors in the operating room (compare figure 2.3). These guiding systems allow clinicians to track and project for example the path of a biopsy needle under visual guidance. Permanent display of target, trajectory, and volume information allows the surgeon for interacting with the surgical plan and monitoring the progress of the procedure [78].

In case that a robot has been registrated to both the patient and to the 3D-image dataset, the robot allows e.g. for positioning tools very accurately at a predefined location or for moving them through a complex path. A teleoperated robot can be programmed to prevent motions into critical regions or only allow motions along a specified direction or surface [44].

A basic problem in stereotaxis is *motion of the target structures during surgery*. Intraoperative guidance is based upon image-data taken prior to the surgical intervention and assumes that there is no movement of the tissue between registration and the surgical act. Therefore, the instrument guidance will be inconsistent if either the patient moves accidentally during surgery, or the actual anatomy of the patient is no longer correctly represented by the 3D-

---

[5] Reprinted from: *Computer Assisted Orthopedic Surgery*, L. P. Nolte, et al. (Ed.), Hogrefe & Huber, 1999.

model. The latter case occurs for example after surgical resection or suction during the intervention and requires the update of the 3D-model *via intraoperative imaging*, if available, or at least the registration process has to be repeated[6]. Alternatively, patient or target motion during surgery can be determined by the tracking system itself, if the patient or the target is provided with markers. Figure 2.4-b shows the mockup of a spine surgery with a LED-marker frame fixed to a vertebra in order to determine spinal motions during surgery. These displacements can be permanently taken into account for correction of the transformation matrix.

2.2 X-ray Fluoroscopy Guided Needle Placement

2.2.1 Common Techniques

X-ray fluoroscopy is a readily available imaging modality in radiology, urology and traumatology. It generates continuous 2D X-ray images and displays them on a television monitor. However, X-ray fluoroscopy provides only poor soft tissue differentiation and applications are therefore restricted either to high-contrast target areas (e.g. lung lesions, bone biopsies) or the operator is depending on the use of contrast agents, e.g. in angiographic procedures [116][133].

In former times radiologists avoided angulated needle insertion - relating to the vertical - because of the lack of angled fluoroscopy or rotating table-tops [172][148]. However, with the development of tiltable fluoroscopes in the 1980s, such as rotational parallelogram fluoroscopes [61] or the commonly employed C-arm fluoroscopes, the radiologist was able to ob-

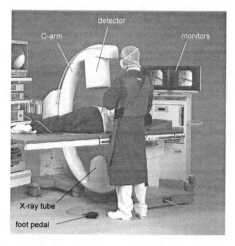

Figure 2.5: Typical configuration of a mobile X-ray C-arm with the X-ray tube on the one side and the image-intensifier (detector) on the opposite side (SIREMOBIL Iso-C, Siemens AG, Germany). The C-shaped arm can be moved and rotated around the patient on the table in order to provide any X-ray view on the monitors. The foot pedal is used to activate radiation.

---

[6] If this re-registration relies on the old 3D-model, one will only get satisfactory results, if the deformations are relatively small or the 3D-model consists of a *deformable model*.

serve the lesion and the needle from arbitrary views. This allowed the radiologist to perform biopsies from various directions and conducting the intervention got much easier (compare figure 2.5).

The *axial aiming technique* became a well established needle guidance method for C-arm fluoroscopy [61][112]. Figure 2.6 demonstrates schematically this alignment technique in case of the puncture of a lung nodule. In a first step the needle tip is positioned at the desired insertion point. Then, the C-arm fluoroscope is rotated till the needle tip (insertion point) and the target are superimposed in the radiographic image, which is shown on the monitor (see figure 2.6-b). Finally, the needle is rotated around the insertion point till its axis is positioned along an X-ray beam, which automatically leads to three-dimensional alignment with the target. After this superimposition of needle hub, needle tip and the lesion in this first view (direction control), the required insertion depth can be observed in a lateral view after rotating the C-arm (depth control). The needle is advanced intermittently one to two centimeter at a time, with a fluoroscopic position check following each advance until the needle has reached the target [172].

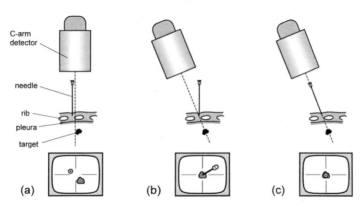

Figure 2.6: *Axial aiming technique* explained by means of a lung nodule biopsy. The lower image border shows the fluoroscopic images in each step. (a) The needle tip is positioned at the insertion point. (b) The X-ray C-arm is moved till the needle tip (insertion point) and the target are superimposed. (c) Now the needle is rotated around the insertion point till its axis is lying in the X-ray central beam, which is automatically leading to 3D-alignment with the target.

Since the needle has to be manipulated in the X-ray field, the operator cannot avoid X-ray exposure. When using X-ray imaging for interventional procedures every effort has to be made to minimize the radiation dose to the patient and the operator. Lead protection is essential while the hands of the operator should never enter the X-ray beam [176]. For this purpose, the use of special needle holders or clamps is often proposed to manipulate and insert the needle in the X-ray beam [27][148]. Alternatively, the needle can be inserted manually in an intermittent approach, where the needle is only manipulated when the fluoroscope is turned off.

Benefits of fluoroscopy are the real-time ability, and due to projective imaging it gives a kind of anatomical overview of a large area, but without direct 3D information. For uncritical target structures or for experienced physicians, fluoroscopic guidance provides a fast and reliable imaging tool [172].

2.2.2 Robots for X-ray Fluoroscopy Guided Needle Placement

The first attempt in using a robot for X-ray fluoroscopy guided needle placement was conducted by Potamianos *et al.* at the Imperial College, London, in 1994 [125][126]. They proposed a robotic system to assist in percutaneous renal needle placement. The system utilized a passive manipulator mounted on the operating room table and guided by a C-arm fluoroscopic unit. The system displayed the access trajectory for the needle on each X-ray image, allowing the surgeon to plan and choose the most appropriate path.

This approach, presented by Potamianos, requires a prior registration of the imaging system (C-arm) to the robot coordinate space. For this purpose the location of the C-arm detector has to be determined with respect to the robot. This is done by using the robotic manipulator as a 3D digitizer with which the location of a set of points provided by a sterile calibration cage attached to the C-arm detector is determined. This registration procedure has to be performed for at least two C-arm orientations which are employed during the intervention.

A similar approach for percutaneous renal procedures has been presented by Cadeddu *et al.* in 1997 at the Johns Hopkins University [25]. However, they used an active robot, called LARS, which was developed jointly by the Johns Hopkins University and IBM Research to aid surgeons in laparoscopic procedures (compare figure 2.7-a; [158]). Imaging was provided by a so-called biplane fluoroscope, which consists of two X-ray C-arms that are oriented under 90°. This allows to acquire simultaneously two perpendicular X-ray views of the patient.

Again, the proposed approach requires a registration of the robot coordinate space to the image space before the intervention can start. For this purpose, the robot's needle holder is provided with a stainless steel sphere [24]. For registration the robot moves the needle holder with the sphere through a series of points spanning the desired target volume (before the patient is placed on the operating table). At each position, the sphere is imaged in both planes by the biplane fluoroscope. The transformation matrix can now be computed from the sphere's location in the images. This approach is more accurate than that of Potamianos, however, the use of a biplane fluoroscope is much more expensive than a simple mobile C-arm system, and it allows only moderate patient access during the intervention.

A very small and compact robot for needle placement has been developed by Stoianovici *et al.* at the Johns Hopkins University in 1998 [154]. The robot is called RCM-PAKY (Remote

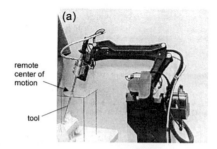
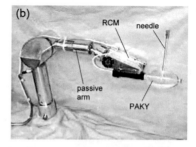

Figure 2.7: (a) The robot LARS has been developed by IBM and the Johns Hopkins University and was initially used to support in laparoscopic procedures.[7] (b) The RCM-PAKY robot, developed by the Johns Hopkins University, has been used for X-ray fluoroscopy guided needle placement via joystick control.

---

[7] In: *Computer Assisted Orthopedic Surgery*, L. P. Nolte, et al. (Ed.), Hogrefe & Huber Publishers, 1999.

Center of Motion, Percutaneous Access to the KidneY). Figure 2.7-b shows the robotic system. It consists of a passive positioning arm and an active remote center of motion manipulator (RCM) with a radiolucent needle driver as end-effector (PAKY, [153]). The RCM allows to rotate the needle around a fixed rotation point (two axes of rotation). The system has been used for percutaneous access to the kidney under joystick control with C-arm fluoroscopy. It does not provide computerized remote control, which is necessary for computer-aided path planning and execution. The basic purpose of this system is to avoid radiation exposure to the hands of the radiologist as in conventional fluoroscopicly guided needle placement.

The intervention starts by manually placing the robot on the patient so that its remote center of motion (needle tip) is located at the needle insertion point on the patient's skin. Then the *axial aiming technique* is performed, as described above. The only difference is that the radiologist orients the needle not by hand around the insertion point, but with the robot via joystick. However, identical is that the radiologist moves the needle according to the X-ray images acquired by the C-arm fluoroscope. The system has been clinically used for several cases at the Johns Hopkins Hospital.

Parallel to these activities of using a robot for fluoroscopicly guided needle placement, there have several similar approaches been developed for orthopedic procedures. Although, in these orthopedic applications mostly a bone drill is used instead of a needle, the principle of image-guided alignment using fluoroscopic images is the same: the bone drill has to be precisely located before drilling and its axis has to be accurately aligned with a certain bony structure, similarly to the alignment of a puncture needle with a lesion.

For example, in 1995, Santos-Munné *et al.* used a Puma-560 robot for X-ray fluoroscopy guided placement of a bone drill for pedicle screw placement [132]. Another orthopedic application has been addressed by Bouazza-Marouf *et al.* in 1995 [17]. This group used a fluoroscopic guided robot for drilling fixation holes into the femur for internal fixation of femoral fractures.

However, all these approaches in the field of orthopedic surgery require the *registration of the robot to the fluoroscopic imaging system* before the surgery can start. In all cases this is done with a calibration frame fixed to the robot's end-effector.

2.3 CT-guided Needle Placement

2.3.1 Common Techniques

CT-guided percutaneous needle biopsy, drainage, and tumor ablation are widely accepted and frequently performed interventions. The ability to localize the needle tip consistently and accurately in an axial plane is a distinct advantage of CT over fluoroscopy or ultrasound [63]. Figure 2.8-a shows a typical computed tomography scanner consisting of the CT gantry and the patient table for moving and positioning the patient within the scanner. Computed tomography is currently the most widely used modality in image-guided therapy [109]. It provides reproducible high-resolution images with good differentiation of soft tissue. Generally, needle placement in the mediastinum, the retroperitoneum and pelvis are in most instances best suited for CT guidance [65]. Furthermore, it is the preferred modality in case that the target area is small (<5 cm), is deeply seated, or is located in critical areas (adjacent to major vessels, nerves, or the hilum of the spleen).

Before beginning the needle puncture, a CT spiral scan of the region of interest is performed for planning of the needle insertion trajectory. An intravenous contrast-medium injection is required at those regions where vascular structures (e.g. the mediastinum) must be demarcated [140]. In the acquired CT image data the most appropriate needle trajectory is chosen, which avoids penetration of vital structures like large blood vessels or nerves. Basically, one can distinguished between:

1. the trajectory is located within the CT scan plane (the whole needle is visible in the image),
2. needle trajectory is oblique to CT scan plane (only cross-sections of the needle are visible)[8].

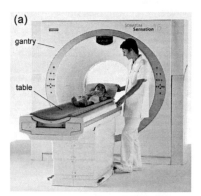
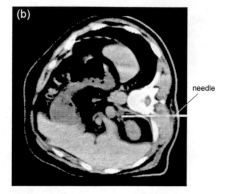

Figure 2.8:   (a) Configuration of a typical CT scanner consisting of the CT gantry and the patient table which can move and position the patient within the scanner (Somatom Sensation 16, Siemens AG, Forchheim, Germany). (b) CT-guided needle puncture of the right adrenal gland. Final control scan after the needle has reached the target. The needle is inserted within the scan plane and therefore completely visible in the image.[9]

Needle placement within the scan plane

There are different needle placement techniques that can be used for fine needle punctures (e.g. for aspiration biopsy) [65]. For example the *single-needle method*, which is a simple but straight method, or the *short cannula coaxial method*. There exist even more CT-guided needle placement techniques, but most of them are very similar to both of these methods, which are described in the following.

With the *single-needle method* the puncture needle is fist placed in the skin and subcutaneous tissue at the defined insertion point. During intermittent CT scanning, incremental adjustments are made by rotating the needle slightly in the scan plane. Once the CT scan shows the correct angulation, the needle is inserted manually to the appropriate depth until the tip is in the proper location.

The *short cannula coaxial method* uses a short guiding cannula (about 4cm) which is placed in the skin and subcutaneous tissues to expedite and optimize needle placement (compare figure 2.9-a). This method is recommended for lung biopsies and small, deeply located masses in the abdomen [65]. The greatest advantage is that the short guidance cannula predetermines the angle prior to the insertion of the puncture needle. Numerous adjustments in the angle can be made without injury to internal structures because the thin puncture needle is not inserted until the correct angle is obtained. Once this is accomplished the needle is inserted

---

8   For small oblique angles it is possible to tilt the CT gantry in order to position the needle trajectory within the CT scan plane (see (a)). However, orientation is getting more demanding for the radiologist.
9   Reprinted from: *Interventionelle Radiologie*, R. W. Günther, et al. (Ed.), Thieme, Stuttgart, 1995.

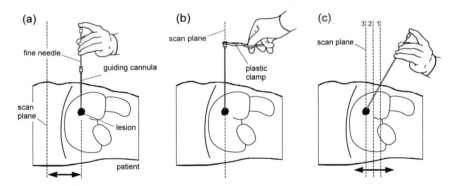

Figure 2.9: Different CT-guided needle placement techniques: (a) Needle placement outside the gantry (with short guiding cannula). The patient is moved into the gantry only for control scans while the intervention itself is performed outside the scanner. (b) Needle placement within the scan plane inside the CT gantry. Depending on whether the operator advances the needle under permanent imaging (compare 'CT-fluoroscopy') or not, he or she has to avoid radiation exposure to the hand, e.g. by using a plastic clamp. (c) Needle placement oblique to scan plane. This approach is quite demanding since the CT image always shows only a cross-section of the needle, which complicates the control of the needle trajectory. By sliding the CT table the needle path can be followed in the image from insertion point to target.

and the only factor to contend with is the distance to the lesion, which can be incrementally adjusted. The cannula can provide access for several repeat biopsies if desired.

Mostly, these needle placement techniques are *performed outside the CT gantry*. The patient is moved into the gantry only for intermittent control scans (see figure 2.8-b) while the intervention itself is performed outside the scanner. The reason for this is the narrow gantry funnel which makes needle placement inside the gantry inconvenient or even impossible. Especially older CT scanners often have a smaller funnel diameter than modern CT scanners and therefore do not admit the patient together with a long, less inserted needle. Furthermore, older scanners have not the technical performance for fast image acquisition, which makes needle advancement inside the gantry a less attractive option anyway. However, this technique of moving the patient in and out of the CT scanner for intermittent imaging can make this sort of needle placement procedure very lengthy.

A very straight and fast method is *needle insertion inside the CT gantry* under direct image control, if the funnel diameter allows for it (see figure 2.9–b). Although, this method might lead to an inconvenient posture for the examiner and the necessity to wear a lead apron, it has significant advantages. The physician has better and direct control of the patient and the whole procedure. The result of needle adjustments are instantaneously visible on a monitor installed beside the CT scanner (compare figure 2.10-a). Moreover, needle bending and drift are directly recognizable. This expedites the procedure significantly and reduces the risk of complications. In this context a very appealing CT scan mode is the so-called *CT-fluoroscopy*, which allows for nearly real-time CT imaging (see section 2.3.2).

Needle placement oblique to scan plane

In case that the skin entry point and the target are *not located in the same CT scan plane*, the needle has to be inserted oblique to the image plane (figure 2.9–c). This technique is required for the sampling of lesions within the pelvis, below the iliac wing, and under the dome of the diaphragm, as well as in trying to avoid certain structures within the abdomen [65].

The problem in this situation is, that the needle and the target cannot be observed at the same time during needle advancement, since both are not located in the same scan plane. Literature is reporting on different methods for oblique needle alignment. For example, some alignment methods are using geometrical approaches to calculate the desired needle orientation. They use the distance between the scan plane of the skin insertion point and the scan plane of the target lesion together with the distance between both planes [11]. The angles for the needle are then calculated according to the Pythagorean theory. However, the basic problem is to execute the calculated angles for needle insertion. In case that the needle is manipulated by hand there have been different *handheld guidance devices* developed. Some of them are consisting of a needle holder with an attached protractor and a simple level indicator, which provide simple but useful support [117]. Another supporting approach for manual needle placement oblique to scan plane is the use of *laser guidance systems*. The calculated angles for needle alignment are adjusted to a calibrated laser system. Consequently, the desired trajectory is indicated by the laser system as the coaxial direction of a fine laser beam [75][55][83] or the intersection line of two, mostly perpendicular laser fan beams [52]. Some of these laser systems are mounted on the CT gantry, others are positioned below the ceiling or on the walls.

2.3.2 CT-fluoroscopy Guided Needle Placement

CT-fluoroscopy is a very appealing CT scan mode for image-guided interventions. With an image acquisition and display rate of up to 8 images per second it provides nearly 'real-time' CT imaging. Thus, CT-fluoroscopy is best suited for needle placement procedures, because it allows precise and fast positioning of the needle with permanent visibility in the CT image. This offers the potential of increased safety when compared with conventional CT guidance [138][54].

CT fluoroscopy was first introduced into clinical practice in 1993 [80][81]. While conventional CT scanning uses one complete gantry rotation (360°) for image reconstruction, CT-fluoroscopy achieves an increased image acquisition rate by repeatedly updating the actual image after each subsequent arc of 120°. This high image sample rate goes to the debit of image quality and the conspicuity of low contrast objects. The reconstructed image matrix has

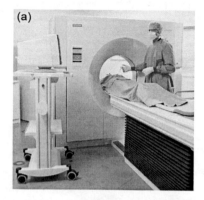
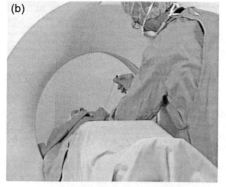

Figure 2.10: CT-fluoroscopy guided puncture. The operator is standing beside the patent and manipulates the needle within the scan plane inside the CT gantry. Scanning is released via foot pedal. Needle advancement is observed in the CT images displayed on an in-room monitor.

only a size of 256×256 pixels instead of 512×512 pixels of a conventional CT scan. However, optimization of algorithms for image reconstruction in CT-fluoroscopy allowed to reduce motion and metal artifacts in the fluoroscopic image and lead to improved low-contrast performance [165][70].

In percutaneous interventions performed with CT-fluoroscopy the operator is standing beside the patient during the procedure (see figure 2.10). The CT images are displayed on in-room monitors next to the patient table directly facing the interventional radiologist. During the intervention the monitor provides immediate feedback about the current needle position, needle drift, or motion of subcutaneous structures and organs.

A drawback of needle placement in the scan plane under direct image guidance is radiation exposure to the operator's hand. To overcome this problem, special *needle holders* or surgical clamps are often used [82][38][146][73]. These tools allow to manipulate and advance the needle while having the hands out of the scan plane (compare figure 2.9–b). However, the needle holder has more or less influence on the image quality, since it may cause artifacts depending on the type of needle holder. There are reports where commercial plastic towel clamps are used [146] or metallic sponge forceps [38][138] to hold the needle. Furthermore, special needle holders have been designed and optimized for CT-fluoroscopy [72][82], basically with a handle on the one side and a plastic adapter for needle fixation on the other.

In case that needle holders are not used, CT-fluoroscopy has to be performed with *'intermittent' CT-fluoroscopy imaging*, where the operator performs manual needle advancement without direct real-time image feedback. After removing his or her hands from the scan plane, the needle position is verified with CT-fluoroscopy, all without leaving the examination room or having to move from the patient's side.

Problems may occur in case that the target is located close to the lungs or the heart. Then, especially small targets may move and leave the CT image plane due to *respiratory motion* or the heart beat [138]. Furthermore, these motions may displace the puncture needle from the scan plane. A remedy in case of respiratory motions are the breath-holding technique or the controlled respiration technique [81]. The *breath-holding technique* may be suitable for lesions that are large or only minimally mobile and when the puncture can be performed within one breath-hold. The technique that may be most useful for mobile lesions, such as those in the lower chest and upper abdomen, is the *controlled-respiration technique*. With this method, the patient has to follow breathing instructions and the operator advances the needle always in the same immobile status of inhalation.

However, the ability to see needle advancement in real-time with CT-fluoroscopy guidance is a distinct advantage over conventional CT guidance. CT-fluoroscopy has allowed to perform procedures that would not have been possible previously [139]. Furthermore, when compared with conventional CT guidance, needle placement time is reduced significantly with CT-fluoroscopy [141][146][151]. Therefore, the drawback of long procedure times, stated by different authors in the past [116][134], cannot be adhered anymore. CT fluoroscopy offers the advantage of target and needle localization with procedure times approaching those of sonographic guidance techniques [144]. Moreover, it has been reported that CT-fluoroscopy guided interventions have an advantage over conventional CT guidance in uncooperative patients because of the ability to visualize a lesion moving in or out of the scan plane in real time [38].

2.3.3 Stereotactic Systems for CT-guided Needle Placement

Many CT-guided percutaneous interventions are performed outside the CT scanner. As described above, this may be because of a narrow gantry funnel or inconvenient patient access, especially within older CT scanners. The physician is conducting e.g. needle placement outside the CT gantry and moves the patient into the gantry only for intermittent control scans.

Therefore, most guiding devices proposed and developed for computed tomography are designed to support in percutaneous interventions outside the CT scanner. Different systems have been developed, which all can be characterized as *stereotactic systems*.

Frame-based *stereotactic systems* in combination with preoperative CT-imaging are commonly used, especially in the field of neurosurgery for localization of brain lesions. However,

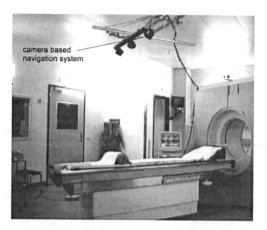

Figure 2.11: CT scanner provided with a 3D optical navigation system mounted below the ceiling. It consists of 3 digital cameras in a line and gives a working area on the table top of 120cm in diameter.[10]

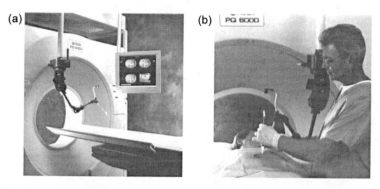

Figure 2.12: The PinPoint™ system, an articulated stereotactic arm for CT-guided needle placement. (a) The stereotactic arm is mounted to the gantry. When the patient has been scanned, an in-room monitor displays the virtual needle path within the 3D anatomy. (b) After determining the optimal needle path, the arm is locked and provides a guide during the ensuing needle placement procedure.[11]

---

[10] Reprinted from: *Computer Assisted Orthopedic Surgery*, L.-P. Nolte, et al. (Ed.), Hogrefer & Huber Publishers, Kirkland, WA, USA, 1999.

[11] Reprinted from: *PinPoint™*, Product Data, Picker International, Inc., USA, 1998.

experiences with stereotactic frames in other parts of the body are relatively limited [115]. Recently introduced optical and electromagnetic navigation systems (frameless stereotactic devices) seem to be of great use in manually performed needle placement procedures (compare section 2.1). Figure 2.11 shows an example of how a CT scanner is provided with an optical navigation system. The infrared 3D digitizer consists of 3 cameras in a line mounted below the ceiling. The system provides an active working area on the table top of 120cm in diameter [76]. All instruments have to be equipped with marker LEDs, as described in section 2.1.

An interesting and commercially available stereotactic arm for CT-guided needle placement is the PinPoint™ system, originally developed and provided by Picker, Inc. The system consists of a gantry-mounted articulated stereotactic arm with a laser positioning attachment at the end-effector (see figure 2.12). The arm's joints are provided with encoders so that the virtual insertion point and needle axis can be computed and displayed on the in-room monitor. After the patient has been scanned, the radiologist simply moves the arm's pointer along the patient's skin while the virtual needle path within the scanned 3D anatomy of the patient is displayed on the monitors in real-time. After the optimal needle path has been determined during this simulation and planning procedure, the entire arm can be locked into place and provides a guide to assist in the placement of the needle or catheter [123].

2.3.4   Robots for CT-guided Needle Placement

Beside the passive stereotactic systems presented above, recently a very progressive and sophisticated approach has been realized: the use of *robotic manipulators* for stereotactic procedures. Many different techniques were developed which integrate robotic guidance of end-effectors with image based stereotactic procedures using a variety of registration techniques.

First experiments on robot guided interventions combined with CT imaging have been performed by Y. S. Kwoh *et al.* in 1985 at the Memorial Hospital in Los Angeles (compare figure 2.13). This group employed a PUMA 200 industrial robot (Unimation Inc.) for precise placement of a surgical probe for *stereotactic neurosurgery* [87][180][45]. After acquisition of preoperative CT-images of the patients head fixed with a special stereotactic head-holder,

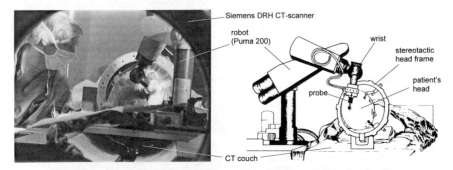

Figure 2.13:   The first robot-supported interventions in medicine were performed in CT-guided stereotactic neurosurgery in 1985. The image shows the surgical setup with a PUMA 200 robot bolted to the patient's couch of a Siemens DRH CT-scanner and a stereotactic head-holder fixed to the patient's head.[12]

---

[12]   On web page: *Robot Assist in Brain Surgery*, NASA Jet Propulsion Laboratory, S. Hayati
http://robotics.jpl.nasa.gov/accomplishments/surgery/surgery.html

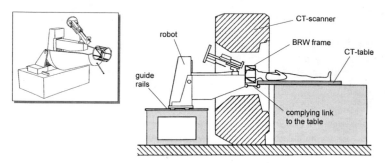

Figure 2.14:   The stereotactic neurosurgical robot MINERVA allows brain interventions directly
inside the CT-gantry. The manipulator and the headframe are part of a sliding unit
which is fixed to the CT-table.[13]

the target coordinates can be determined. Using additional geometrical information provided
by the stereotactic head-frame in the scans, the target-coordinates can be converted into robot-
coordinates, used by the robot for adjacent probe placement. The surgical setup consists of a
stereotactic head-frame fixed to the patient's head on the CT-scanner couch and the PUMA
200 robot (6 *dof* serial-link manipulator) bolted to the same CT couch (see figure 2.13).

Beside the various medical setups using a more or less modified industrial robot, like
Lavallée (Grenoble, France) for stereotactic neurosurgical and orthopedic procedures [91],
some groups initiated the development of entirely new manipulator systems, specialized for a
certain medical application. For example, D. Glauser *et al.* (Swiss Federal Institute of Technology Lausanne, Switzerland) presented in 1989 the concept and design of a stereotactic
robot for neurosurgery in the brain, called MINERVA [56][57] (see figure 2.14). Remarkable
is that the manipulator is operating directly inside the CT-gantry, which allows easy scanning
during surgery. The large system-base is positioned behind a CT-scanner while the manipulator and the stereotactic headframe are part of a sliding unit that is linked to the CT-table. But
again, a stereotactic head frame is used to register the robot and the image space. With this
setup the position of the tool can be easily confirmed during the intervention. However, the
intervention itself is not performed under direct image control. The system is restricted to intracranial procedures. First operations on patients using an aspiration biopsy probe for intracranial cystic lesions were successfully performed in 1993 [48][22].

Similarly, in the early 1990s, Masamune *et al.* developed an isocentric manipulator for
needle placement within the CT scanner [101]. However, the system is smaller than
MINERVA and can be directly installed on the far end of the CT couch (see figure 2.15). As
well as the MINERVA robot, this system is suitable only for intracranial procedures. Registration of the robot to the patient is again performed with a stereotactic head frame. Again, CT
imaging is used only to confirm the position of the needle during the intervention.

A few commercial robotic needle-insertion systems have been developed in the 1990s. One
example is the Neuromate robot, which is a sophisticated manipulator for stereotactic brain
surgery [181]. It was developed by Prof. Benabid, a neurosurgeon from Grenoble, France, and
is now distributed by Integrated Surgical Systems, Inc., USA. The system consists of an articulated arm with 5 axes (see figure 2.16-a). Although this robot has been cleared by FDA for
stereotactic needle punctures, it does not lend itself well to applications inside the CT gantry
due to its relatively large size and heavy build.

---

[13] In Journal: *IEEE Engineering in Medicine and Biology*, vol. 14(3), 1995.

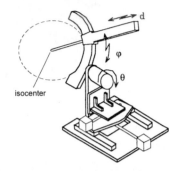

Figure 2.15: Manipulator developed by Masamune *et al.* for stereotactic brain surgery inside a CT scanner. The manipulator is installed on the far end of the CT couch [101].

A new and interesting approach is presented by Nitta *et al.* in 1997 [113][108], who developed a compact master-slave robotic system to perform biopsies in the thorax under real-time CT-guidance inside the CT gantry. The goal of this system is to reduce radiation exposure to the operator during CT-fluoroscopy guided procedures. The system is composed of a robotic arm (modified version of the passive arm 'Point Setter' from Mitaka Kohki Co., Tokyo, Japan) with a biopsy-gun fixed to its end-effector (compare figure 2.16-b). The adjustment of the needle position and the conduction of the biopsy itself is remotely controlled in a mechanical manner by flexible shafts. Although this system is remotely controlled by a human operator, it seems to be the first directly CT-image guided manipulator.

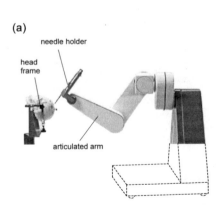
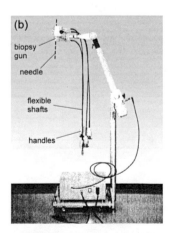

Figure 2.16: (a) The stereotactic neurosurgical robot NEUROMATE, Integrated Surgical Systems Inc., USA. (b) Passive robotic arm for image-guided remotely controlled needle placement inside a CT scanner (manual master-slave system) [113]. A biopsy-gun is attached to the end-effector of this arm and can be manually controlled by flexible shafts.

Another approach has been developed by Susil *et al.* in 1999 [156], who used a very compact robotic manipulator (RCM robot, compare figure 2.7-b) for automatic needle placement inside a CT scanner. The puncture needle is attached together with a special localization frame to the robot's end-effector. The shape of this localization frame is similar to that of a small Brown-Roberts-Wells (BRW) frame (see figure 2.17-a and 2.1-a). In the CT image the frame shows up as a special cross sectional pattern, which is used to compute the pose of the needle relative to the CT image. Although, this method is a remarkable approach of employing stereotactic control with a BRW-frame for interactive CT guidance, in most cases it will cause problems because it needs additional space to move the localization frame together with a needle inside the CT gantry. Due to this fact, a second version of this localization frame has been presented in 2001 with a more compact frame geometry than the initial one (see figure 2.17-b). However, reducing the size of the localization frame results in reduced accuracy in the needle placement procedure.

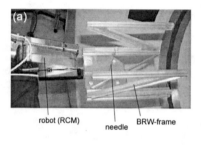 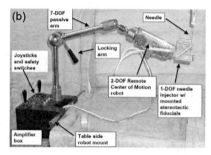

Figure 2.17: (a) Susil's localization frame to perform automatic CT-guided needle placement. The frame is attached together with a puncture needle to a robot's end-effector. The cross-sectional images of the frame show a dot pattern which is used to compute the registration out of a single CT image. (b) Second version of the localization frame with reduced size. However this goes to the debit of precision in the needle placement procedure.

Chapter 3
Robotic Systems in Medicine

Advances in computer and robotics technology allowed manifold integration of robotic technology into health care delivery and medical service. In the last 15 years medical robots have been developed, basically comprising two main areas: rehabilitation robots, and robots in surgery. While first robotic systems with rehabilitative purpose have already been produced in the 1960s, the use of robots in surgery is relatively new. Interest on the employment of robots in the surgical field has been rapidly increasing since the first robotic supported surgical interventions performed by Kwoh in 1985 [87][88]. Since that time robots have been used in several surgical disciplines like neurosurgery, orthopedics or cardiosurgery.

Medical robots have much in common with their industrial counterparts. Actually, the first robots used in the medical field for prototypical setups in medical research centers have been adapted and modified industrial robots, since an entire development of a complex medical robot is very costly and time-consuming. Because of this very close relationship of medical and industrial robots, this chapter starts with a general introduction to computer-based robotic systems, followed by a basic overview of manipulator kinematics and control.

The author sees great use in analyzing already existing robotic systems in the medical field, in terms of geometry, kinematics, control, or safety concepts, before starting with the development of a new needle guiding robot for radiology. Furthermore, the way of integrating the robot into the clinical workflow is an important issue and an important pre-condition for a broad acceptance in the medical field. Therefore, the subsequent sections deal with various robotic applications in health care settings and give an overview of manipulator designs applied in medicine.

3.1 Computer-based Robotic Manipulators: A Short Survey

In recent years, the use of robotic manipulators has increased rapidly, particularly in such industrial settings where the need for greater productivity and tighter quality control has been of significant importance because of intense international competition. Industrial robots became famous in applications that require repetitive tasks over long periods of time, operations in hazardous environments (nuclear radiation, under water, space exploration), and precision work with a high degree of reliability.

The first robot manipulators came up with the beginning of the nuclear age and were known as teleoperators [142]. These devices allowed a human operator to perform tasks from a safe distance, such as handling of radioactive materials, in a master-slave fashion. The first electric-powered teleoperator was developed by Argonne National Laboratory in 1947 [142][68]. Whereas these early teleoperators were controlled by a human and without force feedback or other sensory information, the programmable computer-controlled manipulators as used today had their advent in the numerical-controlled machines introduced in the early 1950s. The M.I.T. demonstrated the first servo-controlled numerical milling machine in 1953. These early numerically controlled systems were programmed by means of punched paper tapes, providing the digital information of the robot's movements for example. However, first commercial robots became available not before the mid-1970s [64].

Since the early 1970s robotic research has been extended to the development of sensors to provide information feedback to the robot control system about the task being performed.

Beside improved *internal sensors* like different types of joint encoders, force and torque sensors, extensive research has been devoted to *external robotic sensors.* In 1966 the Stanford's Artificial Intelligence Laboratory developed a manipulator robot with ears, eyes, and hands (microphones, TV cameras, and manipulators) which could recognize spoken commands and respond accordingly [103]. In the following years extensive work has been done on external sensors like vision systems, tactile sensors, sonar systems (providing a measure of distance), and speech recognition [142].

Early applications of robotic manipulators have been used in hazardous working environments, as for spray painting [46], spot-welding applications [49], or for handling parts for a stamping press or forging machine [84]. Furthermore, in the 1970s manipulators have been employed for installation of underwater power cables using mechanical arms attached to a submersible vehicle [142].

Much of this sophisticated technology, originally developed for industrial robotic applications, has been adopted and integrated for medical robotic systems working in a clinical surrounding.

3.1.1 Components of a Robotic System

This section describes the configuration of general-purpose robotic manipulators, considering their degrees of freedom, work envelopes, end-effectors, and types of actuators. These manipulators are integrated in a robotic system that uses a robot controller, external sensors, and a host computer (see figure 3.1). The host computer interprets operator inputs and sends low-level commands to the *robot control* which is controlling the manipulator motions. The robot controller samples *external sensors* that detect changes in the robot's environment.

The robotic *manipulator* is the principal component of the robotic system. The manipulator, in turn, is composed of a *base, links, actuators,* and an *end-effector.* The base usually is fixed to a supporting surface and provides a fixed reference system, which helps in robot programming. The manipulator links are its movable segments, connected by joints in between.

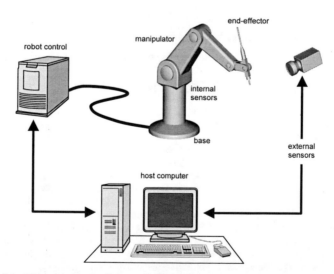

Figure 3.1: Major components of a general robotic mechanical system consisting of the robotic manipulator and end-effector, the control system, internal and external sensors, and a host computer.

Depending on the combination of joints, revolute or prismatic, manipulators are *Cartesian, cylindrical, spherical, scara* or *revolute* [23]. The number of manipulator joints corresponds to its number of degrees of freedom (*dof*). The manipulator wrist allows the manipulator to orient the end-effector in space. The work envelope of a manipulator is the 3D space wherein the robot can reach and perform its programmed task.

Further components of the manipulator are its actuators and internal sensors. *Actuators* are used to move the manipulator to the desired position and orientation whereas internal sensors are measuring joint positions for precise motion control. In a medical environment the most appropriate actuators are *electric drives*, since they are clean, silent, easy to control, and easy to install and maintain [23].

A manipulator interacts with the task it has to complete through the end-effector mounted on its wrist. Depending on their geometry, end-effectors may *be grippers, specialized tools*, or *dexterous hands*. Typical tasks may require drilling, welding, or picking up objects. End-effectors may have their own actuators and sensors, such as optical, force or touch sensors to obtain sensory information during the operation.

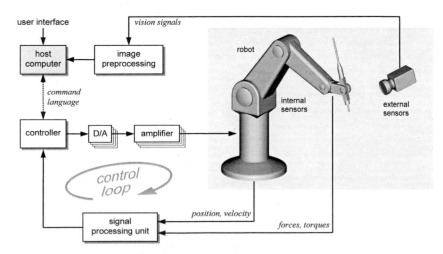

Figure 3.2: Diagram of the basic organization of a robotic system, consisting of the manipulator with internal and external sensors, the robot control loop with several signal processing units, and the host computer for robot supervision and robot programming. Note, that every manipulator joint needs its own low-level control loop.

With the employment of *external sensors*, the system capability could get increased in a way, that tasks were performed successfully even in unstructured and - for the robot - unknown environments. External sensors are needed to provide information on the work space and to allow the robot to adapt to changes. For this purpose *vision sensors* are increasingly used, especially for tracking of objects or for analyzing the workspace. Vision sensors are solid-state cameras mounted above the work envelope (fixed) or on the manipulator arm itself (mobile) [35]. The camera images are received by the image preprocessing unit and sent to the host computer, where they are analyzed and used for ensuing motion commands which are sent to the manipulator controller (see figure 3.2). The visual information of the camera is sometimes directly processed in the control loop, leading to a so-called *visual servoing* feed-back control, which is extensively discussed in chapter 6.

The *robot control* is a basic component of a robotic system. Figure 3.2 shows the organi-zation of a robotic system and a scheme of its control. The desired task is described to the controller by means of software programs on the host-computer. There are different types of high-level control to define the task on the host computer: manual control, off-line control, and task-level control (see e.g. [23]). The low-level manipulator control involves several con-trol loops, each controlling a manipulator joint. The actual position and velocity information of the end-effector and links are provided by internal sensors, e.g. motor encoders or ta-chometers. Every control loop than compares the current positions of the link with the desired position given by the host computer and derives new motion parameters for the manipulator. The control feedback loop is activated till the desired position and orientation of the end-effector is achieved [8]. Additional information can be provided by external sensors - mostly vision sensors - for higher precision and reliability in execution of the task.

For the successful performance of a task, accuracy and repeatability of a manipulator are crucial. Accuracy results from the *spatial resolution*, which is defined as the smallest incre-ment of motion achieved by the manipulator's end-effector. It is the combined resolution in-cluding the *control resolution* (the smallest increment that the controller can control by means of D/A or A/D converting) and the *motion inaccuracies* (resulting from e.g. vibrations or elastic deformations) [142]. *Repeatability* refers to the error with which a robot returns to a previously taught or commanded point. In general repeatability is better than accuracy, and is related to joint servo performance [35][23].

3.1.2 Manipulator Kinematics

It is clear that the location and orientation of the end-effector is the result of the collective effect of translation and rotation of each joint of the manipulator chain of links. The *pose* of e.g. the end-effector can be described in terms of the *position* and *orientation* of a coordinate frame attached to the end-effector (see figure 3.3). In order to describe the robot's physical structure and the task that is to be performed, several coordinate systems are employed. While interactions between the robot and its environment are commonly described in a *world coor-dinate system*, the task itself is defined meaningfully in relations between the *end-effector-* and the *base coordinate frame*. To describe the relative motion between the links of the robot, it is appropriate to attach *link coordinate systems* to every link. Typical robots are serial-link manipulators which can be described as open-loop articulative chain of links connected in series by either prismatic (sliding) or revolute (rotary) joints. Thus, each joint has one degree of freedom, either translational or rotational [35].

Translational and rotational motion of the end-effector can be described mathematically by means of a 4×4 homogenous transformation matrix ${}^B H_E$, which is representing the pose (po-sition and orientation) of the end-effector frame E with respect to base frame B:

$$
{}^B H_E = {}^B_E[RT] = \begin{bmatrix} {}^B R_E & {}^B T_E \\ 0 & 1 \end{bmatrix} = \begin{bmatrix} R_{xx} & R_{xy} & R_{xz} & T_x \\ R_{yx} & R_{yy} & R_{yz} & T_y \\ R_{zx} & R_{zy} & R_{zz} & T_z \\ 0 & 0 & 0 & 1 \end{bmatrix}. \tag{3.1}
$$

The transformation matrix ${}^B H_E$ consists of the rotation matrix ${}^B R_E$, representing the orienta-tion of frame E with respect to frame B, and the translation vector ${}^B T_E = [T_x \, T_y \, T_z]^T$ from frame B to E. The homogenous notation includes both the rotational and the translational informa-tion in a single matrix. This notation is advantageous because combinations of several rota-

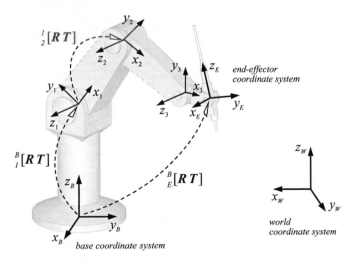

Figure 3.3: Schematic of a serial-link manipulator with all its coordinate frames. The whole kinematic chain can be expressed in "direct kinematics" as a linear transformation $_E^B[RT]$ representing a translation T and a rotation R to define the pose of the end-effector frame $[x_E \ y_E \ z_E]$ in base coordinates $[x_B \ y_B \ z_B]$. To describe the robot with respect to its environment, commonly a world coordinate frame $[x_W \ y_W \ z_W]$ is used.

tions and translations, of each joint for example, can simply multiplied to one resulting transformation matrix:

$$_E^B[RT] = {}_1^B[RT]\,{}_2^1[RT]\ldots{}_E^{n-1}[RT].\qquad(3.2)$$

This approach is called *direct kinematics*, when multiplying each homogenous joint transformation to obtain the resulting pose of the end-effector in relation to the base coordinate frame. On the other hand, the *inverse kinematics* problem for robotic manipulators involves finding the homogenous joint transformations $_i^{i-1}[RT]$ for each joint $i = 1, 2, \ldots n$ for a certain end-effector pose. This approach is important in order to find the resulting joint motions, if a certain end-effector displacement has to be performed.

This brief introduction to the basics of robotic technology and notations should be sufficient for the comprehension of the actual thesis and for the development of a novel micro-robot for image-guided needle placement, which is presented in chapter 5. For further information about these topics it is referred to special robotic literature (see [8][26][64][142]).

3.2 Robots in Medicine

Although the general functions of a surgical robot are quite similar to that of a surgeon, the properties that result are quite different. The general intention is that such robots should assist and support the surgeon in difficult interventions, not replace the physician. The use of robotic systems in medicine has already started in the 1960s in the field of robotic prosthetic [39]. However, most attention has been attracted by the first successful applications of robots for surgical procedures in the early 1990s. Nowadays, medical robots have achieved widespread dissemination in health care settings. The four main areas of medical application are: *surgery*, support for *disabled* people, *rehabilitation*, and *medical service* (compare figure 3.4). The following sections summarize the results of a literature review in medical robots and their applications. Focus has been set on manipulators used for *surgical interventions*.

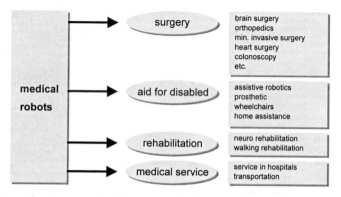

Figure 3.4:   Overview of the different medical fields where robotic manipulators are in use.

Before starting with the presentation of various medical robotic systems, an important issue shall be discussed: the *safety* of medical robots. Unlike most industrial manipulators, medical robots commonly operate in close physical contact with humans. This requires extremely safe, stable and compliant operation of the robot control system and the manipulator itself. The difficulty is that medical robots do not have generally agreed safety recommendations [42][44]. Industrial robots are required to operate inside a cage, away from people, and are only powered up when all personnel are excluded. This is clearly inappropriate for surgical robots in the operating room where robots are working very closely together with surgeons and the patient. Safety aspects are the main reason why medical robots are generally moving very slowly.

As already mentioned in section 2.3.4, the use of robots in surgery has its roots in the field of *stereotactic neurosurgery*, when Y. S. Kwoh employed a PUMA 200 industrial robot for precise placement of a surgical probe at the Memorial Hospital in Los Angeles in 1985 [87][180].

The pioneer of *orthopedic robotics* is undoubtedly R. H. Taylor, who developed in the late 1980s at IBM a robotic system for cementless hip replacement surgery. The system is a 'SCARA' style of manipulator based on a modified IBM 7576 industrial robot (see figure 3.5). The end-effector is provided with an additional pitch axis, a 6-*dof* force sensor, and a high-speed cutting tool. The rotating cutter reams out the proximal femur for insertion of the femoral stem of a prosthetic implant for total hip replacement. The system consists of two main components: the preoperative image-guided planning workstation ORTHODOC™ and the surgical manipulator ROBODOC™ (now trademarks of Integrated Surgical Systems, Inc.). Using preoperative CT-imaging, the surgeon can create a preoperative plan with the ORTHO-DOC™ software, which will afterwards be executed by the ROBODOC™ system. The robot-to-patient registration is achieved by three titanium 'locator pins' (bone screws) placed in the femur prior to the CT scan. Before starting the surgery, the surgeon guides the manipulator arm to each locator pin for automatic pin position localization [157][159]. In 1992 a first clinical trial was permitted by the US Food and Drug Administration (FDA) [106][16].

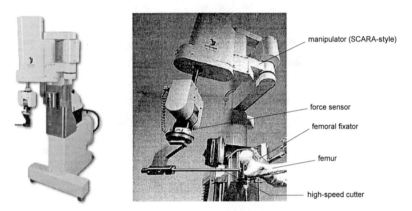

manipulator (SCARA-style)

force sensor

femoral fixator

femur

high-speed cutter

Figure 3.5: The ROBODOC system for cementless total hip replacement surgery. With a height of more than 2 meters it is one of the largest surgical robots.

At the same time B. Davies at the Imperial College in London presented a robotic system for the removal of *prostatic tissue* [40]. Initially, he adopted an industrial robot (PUMA 560, Unimation Inc.) supported with an additional manipulator mounted on to the PUMA robot, resulting in an 8-*dof* system. Davies realized that the use of an industrial robot with many *dof* and designed to have a large envelope of motions would be intrinsically less safe than that of a *special purpose manipulator* whose motions and forces were designed specifically for the desired task [44]. Taking these considerations into account, Davies developed a second-generation prostate robot ("Probot") mounted as an 'end-effector' on a large floor-standing counterbalanced passive manipulator arm. The physical design of this end-effector constrains the robot motions to a well defined limited area of operation (see figure 3.6). The resecto-scope is moved around a remote center of motion, which is geometrically defined by its de-sign (center of arc). In the event of failure, an anthropomorphic robot capable of reaching a large-volume space, could fly off in any direction, causing damage to patients, surgeons, and other bystanders [42].

Beside the various medical setups using a more or less modified industrial robot, like Lavallée [91] or Santos-Munné [132], some groups followed the idea of Davies and initiated the development of entirely new manipulator systems, specialized for a certain medical appli-cation which only use as much *dof* than necessary. For example, C. W. Burckhard and D. Glauser (Swiss Federal Institute of Technology Lausanne, Switzerland) presented in 1989 the concept and design of the stereotactic robot MINERVA for neurosurgery in the brain [56]. The large system-base is positioned behind a CT-scanner while the manipulator and the stereotactic headframe are part of a sliding unit that is linked to the CT-table (compare figure

Figure 3.6: The manual version of the "Probot" end-effector. This special fame requires only 2-
dof for placing the resectoscope. The resectoscope is carried on a circular arc that, in
turn, is mounted on a ring which is free to rotate through 360°.[14]

2.14). First operations on patients using an aspiration biopsy probe for intracranial cystic lesions were successfully performed in 1993 [48][22].

In 1994 the US company ComputerMotion Inc. presented the first commercially available
medical robot in the field of *minimally invasive surgery* [130]. The purpose of this robot
called AESOP™ (Automated Endoscope System for Optimal Positioning) is to hold and position the endoscopic camera, while the surgeon manipulates the instruments manually for performing the surgery (see figure 3.7-a). Camera holding is normally done by a surgical assistant or nurse. The AESOP robot eliminates the need for an additional person at the operating

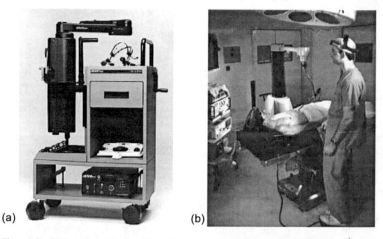

(a)   (b)

Figure 3.7: Two commercial robots for endoscopic camera control: (a) AESOP: for surgery the
manipulator arm is removed from the carriage and attached to the rail of the surgical
table. (b) EndoAssist (current product name for EndoSista): floor-standing system;
the surgeon is controlling the camera position via motion sensors in his headband.

---

[14] In: *Computer-Integrated Surgery - Technology and Clinical Applications*,
R. H. Taylor et al. (Ed.), The MIT Press, Cambridge, MA, USA, 1996.

table and may therefore lead to cost savings. The surgeon controls the robot by means of either a foot or hand controller and is enabled for so-called minimally invasive "solo-surgery". Adjacent developments provided the system with increased precision and additional features like voice-control (AESOP 3000) [105].

At the same time the British company Armstrong Projects Ltd. developed the EndoSista robot, also for endoscopic camera manipulation [51]. This is a floor-standing system with a remarkable type of camera control: the surgeon is wearing a special headband with magnetic sensors measuring the surgeons head movements. The surgeon simply turns the head in that direction he or she wants to move the camera (while pressing a footswitch) and the robot moves the endoscopic camera respectively (see figure 3.7-b).

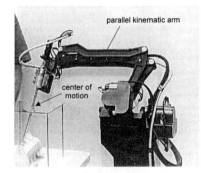
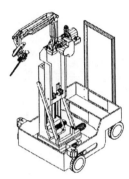

Figure 3.8:   Parallel link manipulator LARS for laparoscopic interventions, developed by IBM and the Johns Hopkins University.[15]

Beside these commercially available manipulators several research groups have been developing robotic systems for endoscopic surgery as well. For example the LARS robot (Laparoscopic Assistant Robot System), also developed by R. H. Taylor in a collaboration between IBM and the Johns Hopkins University Medical School in the early 1990s [158]. The robot has a remote center of motion, which is defined by the parallel linkage structure of the robot arm (see figure 3.8). During surgery this point is identical with the endoscope insertion point in the patients abdomen. The robot motion can be controlled with a small joystick device clipped onto an endoscopic instrument. With the end-effector equipped with a 3D-force sensor the surgeon can directly move the endoscope by pushing it softly into a certain direction. Additionally, the system is capable of video image processing for simple automatic image-guided instrument placement [159].

The above described robots for minimally invasive surgery are basically used for camera holding and movement, while the surgeon is still standing beside the patient to perform the intervention manually in a conventional manner. However, in the ensuing years minimally invasive robotic systems have been developed with the purpose of grasping, cutting or suturing tissue in an teleoperated way. For example in 1997, ComputerMotion Inc. presented the ZEUS™ robot [129] (figure 3.9-a), which is an extended and telemanipulated version of the AESOP system. At the same time, a quite similar system, named daVinvi™, was developed by IntuitiveSurgical Inc. [145][96] (figure 3.9-b). Both systems consist of two basic compo-

---

[15] In: *Computer Assisted Orthopedic Surgery*, L. P. Nolte, et al. (Ed.), Hogrefe & Huber Publishers, 1999.

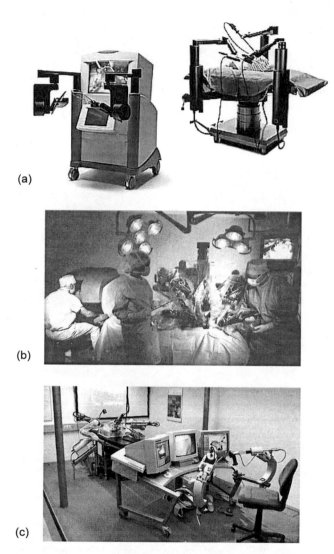

Figure 3.9: Three teleoperated robotic systems for endoscopic surgery. The surgeon is sitting at a console provided with a vision system and special handles for remote controlled manipulation of the robot. (a) ZEUS: mobile console with instrument-handles, three AESOP robots attached to patient table. (b) daVinci: console with instrument-handles and excellent 3D-view, large robot base with 3 manipulator arms (c) ARTEMIS: large master-handles, no 3D-view.[16]

[16] In: Nachrichten - Forschungszentrum Karlsruhe,
Research Centre Karlsruhe GmbH, vol. 32(1), 2000.

nents: (i) the surgeon's viewing and control console, and (ii) the manipulator system with the surgical instruments. During surgery the physician is seated at the console provided with a 3D-vision system and special instrument handles for remotely controlled manipulation of the surgical instruments. In both systems the handle-movements are scaled down and tremor filtered for microsurgical procedures. The major difference between both systems is the articulated instrument tip in the daVinci system. While ZEUS is using straight stiff instruments, those employed in the daVinvi system are provided with additional joints in the instrument's wrist [29]. The main applications of both systems are minimally invasive cardiac interventions. They allow to eliminate the need for the large incision through the sternum and the heart-lung machine, which causes much less trauma on the patient than conventional open heart surgery. The author had the chance to take part in a demonstration of the DaVinci system (on a phantom) in the Herzklinik Munich, Germany, in July 2000. He could assure himself of the excellent 3D-view and the smooth and flexible transmission of handle-to-instrument movements supported with force-feedback.

A related telemanipulated system for minimally invasive surgery, named ARTEMIS, was presented by the Research Centre of Karlsuhe, Germnay, in 1995 [162][21]. The system consists of the control unit with monitors and two large master-handles on the one side and the two slave-manipulators attached to the patient table on the other (see figure 3.9-c). This system had a more scientific purpose and was never tested in a clinical environment on patients.

Another German research group at the Frauenhofer Institute IPA, Stuttgart, presented a teleoperated hexapod robot for minimally invasive brain surgery in 1997 [167]. Again the system consists of a control unit and the hexapod manipulator attached to a C-ram base (see figure 3.10-a). The control unit is a large movable chair ('operating cockpit') with display and integrated joysticks in the two armrests, which allow for interactive control of the endoscope and the instruments. The chair movements are linked to the movement of the endoscope and should give the surgeon the illusion of being located on the tip of the endoscope tip [161]. The manipulator with its parallel-kinematic is a commercially available hexapod robot (Physik Instrumente GmbH, Waldbronn, Germany) with an absolute position accuracy of better than $20\,\mu m$. In 1998, these activities were adopted by the company URS GmbH, Parchim, Germany, a spin-off company of the Frauenhofer Institute IPA. In 2000, URS came with their new hexapod system on the market (see figure 3.10-b).

Another orthopedic robot, named CASPAR™, was presented in 1998. The purpose of this

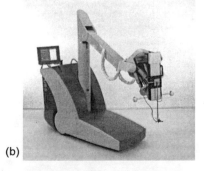

(a)   (b)

Figure 3.10:   The hexapod robot for minimally invasive neurosurgery, developed by the Frauenhofer Institute IPA, Stuttgart, Germany. (a) The first prototype presented in 1997.
(b) After further developments the sin-off company URS GmbH, Parchim, Germany, came with a new mobile hexapod system on the market.

manipulator, developed by the German company OrtoMAQUET GmbH, Rastatt, is hip- and knee-replacement surgery [122]. The manipulator is mounted on a mobile base of a MAQUET operating table (see figure 3.11). CASPAR is smaller and more compact than the ROBODOC system, but their workflow, the registration procedures, and the control during surgery is very similar.

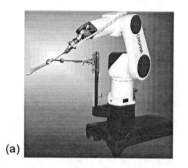

(a)   (b)

Figure 3.11:   (a) The orthopedic robot CASPAR, developed by the company OrtoMAQUET GmbH, Rastatt, Germany. It is a modified industrial robot for clean-room applications.
(b) Computer integrated OR at the Charité-Virchow-Hospital, Berlin, Germany. The setup consists of a PUMA560, a ceiling mounted robot SurgiScope, and a mobile CT-scanner.

An interesting approach for a surgical manipulator was realized by the Surgical Robotics Lab (SRL) at the Charité-Virchow-Hospital, Berlin, Germany. They employed a medical robot, the SurgiScope (Elekta Instruments, Inc./DeeMed), which already had the FDA approval for placement of a surgical microscope (see figure 3.12-b), and modified the end-effector adaptor for the fixation of surgical instruments. Figure 3.11-b shows the OR-setup consisting of a PUMA 560 robot, the ceiling mounted parallel-robot SurgiScope, and a mobile CT-scanner (Philips Medical Systems Inc.). The medical purpose of this robotic setup is maxillofacial surgery and the therapy of hyperthermia [98][97].

The SurgiScope leads to the last group of surgical robots presented in this section: robotic carrier systems for surgical microscopes (see figure 3.12). The most popular examples are the MKM system, developed by Carl Zeiss AG, Germany, and the already mentioned SurgiScope. Both systems are used for intraoperative navigation [60].

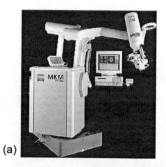
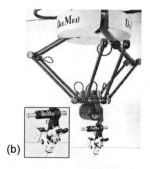

(a)   (b)

Figure 3.12:   Two robotic systems for surgical microscopy. (a) MKM, (b) SurgiScope

Beside all these presented products and the early research activities leading to them, there are various other research groups in the field of medical robotics all in the world, which to mention would go beyond the scope. But this substantial overview of the most important medical applications for robotic systems and their implementation concepts leads to a classification of different types of surgical robots:

automatic (predefined) positioning tool: The robot moves a tool precisely to a preoperatively defined pose. Early applications of robots in medicine, especially stereotactic surgery (compare CAS, e.g. Kwoh [87]).
teleoperated robots: Master-slave systems. The surgeon is sitting at a console with monitors and moves the end-effectors of the robot remote controlled via force-feedback joysticks or special master-handles. The robots are not working autonomously.
fully automated robotic system: Robots are performing preoperatively planned tasks. Mostly used for drilling/milling of bones in joint replacement surgery in orthopedics. Procedure requires the registration of the preoperative plan/images to the patient before starting the surgery. Then the robot executes a completely pre-programmed trajectory sequence.
navigated interactive robots: Robots are 'tool support systems' that carry, guide and move surgical instruments. Typically they are navigated and use similar registration methods as the fully automated robots. The end-effector is moved manually by the physician via joysticks or buttons at the end-effector tool itself. The intervention is often supported by navigation aids, direct imaging (video or microscopic), or predefined constrained movement areas.
automatic image-guided system: Image-guided robots move and orient themselves by real-time images acquired during the intervention. An image-based control analyses the video, X-ray, MR or ultrasound images and moves the end-effector with regard to certain image features extracted out of the images.

Table 3.1:   A classification of the various types of robotic systems in surgery, starting with the early applications like automatic tool positioning, to the automated image-guided systems.

Many scientists, industrialists, and physicians perceive the area of surgery as a very promising field of application for robotics [39]. Beside the often existing suspicion of robots in surgery, there exist many good reasons for their employment in the operating room. The advantages can be summarized as following:

- robots can apply well-defined forces with high positional accuracy (precise trajectories),
- equipped with sensors, the robot can measure parameters during surgery (e.g. forces),
- the execution of the robot's task can be exactly planned and simulated before surgery,
- the task can be performed remote-controlled with scaled and tremor-filtered motions,
- a teleoperated robot can be programmed to prevent motions into critical predefined areas,
- a robot can react much faster on certain incidents than a human and can coordinate complex processes even together with other machines in the OR within milliseconds,
- robots do not care about X-ray radiation for image acquisition during the intervention,
- robots do not show any tremor and do not get tired holding the tool for a long time.

However, except for a few cases probably, robots in the future will not execute operations fully autonomously but will increasingly support surgeons in the OR to achieve optimal surgical results.

Chapter 4
Market Analysis

The last two chapters demonstrated, that innovative radiological and surgical applications such as image-guided procedures, robotic interventions or surgical navigation are offered by the research field and partly already exist in the medical theater. Yet, despite such technological advances, skepticism exists among physicians, scientists and governmental agencies as fears about affordability, accuracy, and optimum performance come to mind. Will these technical innovations really solve clinical problems, and will they improve patient outcomes in a cost-effective manner?

Success of a robotic system on the medical health-care market strongly depends on these factors and requires an objective review of problems and needs of the medical specialties that are focussed for this robotic tool. Furthermore, it is essential for a broad acceptance of a medical robot, that its high investment costs will be compensated or even passed by cost-savings resulting from the use of the robot.

Thus, before starting the development of an image-based *needle guiding robot* for medical applications, the author realized the need to collect information about

- how is the physician performing image-guided needle placement?
- which would be the key applications for such a system?
- how many interventions are performed per year?
- what would be the desired medical improvement by such a system?
- which technical requirements would this system have to meet?

In order to get answers to these questions, the author organized and performed several physician interviews in July and August 1998 [7]. The results of these interviews lead to the definition of a *clinical requirement specification* for the robotic system which is presented at the end of this chapter. From these clinical specification, a *technical requirement specification* has been created in which all hardware and software requirements were defined to develop a prototype of an image-guided needle-placement robot (see chapter 5).

Initially, the interviews have been confined to the field of radiology, focussed on interventional radiologists. However, the first meetings with several radiologists turned out that common fluoroscopy is mainly used in radiology for angiographic interventions. But most needle placement procedures are increasingly performed under CT-guidance than under X-ray fluoroscopy. Due to these initial information the interview group was extended to further medical disciplines frequently working with X-ray fluoroscopy, like traumatology and orthopedics. Additionally, all radiologists have been asked about the need of a guiding system for CT-interventions.

Table 4.1 shows the list of all interviewed physicians, their medical discipline and the medical sites where they are working.

interviewed physicians	medical site
8 radiologists	Knappschaftskrankenhaus Bochum Zentralklinikum Augsburg Klinikum Süd, Nürnberg Städtische Kliniken Dortmund Uniklink, Erlangen Mühlheim, Universität Witten/Herdecke St. Josefs-Hospital Dortmund Klinikum Wiesbaden
3 / traumatology	St. Katharinen-Krankenhaus Frankfurt Chirurgische Klinik, Erlangen Zentralklinikum Bamberg
2 orthopedists	Klinik Lindenlohe, Schwandorf Rummelsberger Anstalten, Lindenlohe
3 urologists	St. Katharinen-Krankenhaus Frankfurt Urologische Klinik, Erlangen Urologische Klinik, Mainz
1 general surgeon	Chirurgische Klinik, Erlangen
1 / pneumonology	Klinikum Süd, Nürnberg

Table 4.1:   List of all 18 interviewed physicians, their disciplines and medical sites.

4.1   Summary of Interview Results

Before the acquired data is presented in detail, this section wants to summarize the basic statements and valuations of all interviewed physicians. Figure 4.1 and 4.3 show the principle attitude of all physicians towards the proposed image based guiding system. All statements about usefulness and sales opportunity must be seen in regard to the respective medical filed of the interviewee. Only radiologists have been asked about the need of a *guiding system for CT-fluoroscopy*. All assessments made by the interviewees are categorized in

"yes"	Interviewee sees useful applications for an image based guiding system in his/her field and suggested to use such a system.
"skeptical"	Guiding system could probably be used for several applications, but interviewee sees problems of widespread acceptance in the medical field.
"no"	Interviewee sees no use or benefit in the proposed guiding system, even if it principally could be employed for some applications.

4.1.1   Guiding-system for X-ray Fluoroscopy

Figure 4.1 shows the low acceptance of an X-ray fluoroscopic guiding-system in the group of radiologists. The same for the pneumonologist and the general surgeon. The need for such a guiding-system under the interviewed urologists is varying. While one urologists is optimistic, the other explained that he sees no chance for such a system on the German/European market. However, he referred to the US market where X-ray fluoroscopic imaging is widely employed in urology, thus the proposed guiding system may receive a larger acceptance in US than in Europe.

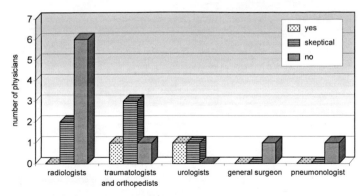

Figure 4.1: Assessment of acceptance of a guiding-system for X-ray fluoroscopy in different medical disciplines.

All traumatologists and orthopedists mentioned various guiding-problems during their surgeries, like precise drilling or screw fixations. For these procedures usually X-ray imaging is employed to control the intervention. Therefore in this group most physicians would appreciate to have a guiding-tool in certain applications for precise placement of screws or drill holes.

medical field	applications	
radiology	lung applications	puncture of pulmonary densities or tumors
	bone applications[17]	puncture of bone tumors general bone punctures (no 3D-bones)
	neurology applications	spine puncture nerve root block after myelography
	puncture of other organs	liver and biliary system with contrast media kidney (with contrast media)
traumatology and orthopedics		placement and distal fixation of medullary nails placement and guidance of Kirschner-wires placement of single screws (without implants) puncture of bone tumors spine puncture
urology		kidney puncture puncture of the biliary system
general surgery	interviewee sees no useful applications in his field	
pneumonology		puncture of peripheral lung tumors

Table 4.2: Summary of statements given by the different groups of physicians with regard to possible applications of a guiding system under *X-ray fluoroscopy*.

Table 4.2 shows a survey of the basic statements given by the different groups of physicians with regard to possible applications of a guiding system under X-ray fluoroscopy. Referring to [7], these applications could be summarized with regard to frequency and their estimated importance as shown in figure 4.2.

---

[17] This is with regard to "simple shaped" bones. Punctures of bones with a complex 3D structure, like pelvis or shoulder, are commonly preferred to be done under CT-guidance, according to the interviewees.

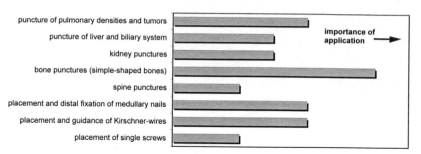

Figure 4.2: Relevance of possible applications with a guiding system under *X-ray fluoroscopy*, according to the interviewees.

4.1.2 Guiding-system for CT-fluoroscopy

Beside the discussion on a guiding-system under traditional X-ray fluoroscopy, all radiologists have been asked about the usefulness of a guiding system inside a CT-scanner, used under *CT-fluoroscopy imaging*. As shown in figure 4.3, the majority of radiologists see a significant advantage of such a guiding system used inside the CT-gantry directly under image-guidance.

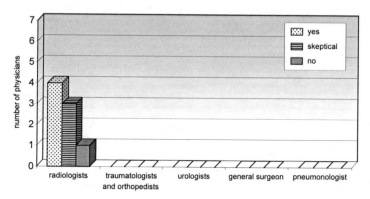

Figure 4.3: Assessment of acceptance of a guiding-system for CT-fluoroscopy in the group of interviewed radiologists (the other groups have not been asked about this issue).

4.2 The Interviews

In this section further information about the various image-guided interventions, their workflow and resulting functional requirements are presented. All these data have been gathered from the interviews and are presented separately in the following two subsections dealing with X-ray and CT-guided procedures.

4.2.1 A Guiding-system for X-ray Fluoroscopy

Radiology

The majority of interviewed radiologists explained that nowadays X-ray fluoroscopy is only rarely used for guiding needle placement procedures. This is basically because of the inadequate soft tissue contrast in X-ray fluoroscopy images, which is leading not only to the problem of indistinct representation of the target structure, but to inadequate visibility of vital structures like huge vessels or nerves, that have to be avoided on the way to the target. Structures fairly visible under X-ray fluoroscopy are tumors and calcificated tissue in the lung. Small tumors or metastases, especially in the abdomen, are usually not visible. Before the 1990s, when X-ray fluoroscopy has been the imaging modality of choice for most procedures in interventional radiology, the tumor could only approximately be aimed. In cases of pulmonary densities, for example, the tumor might be lightly visible under X-ray fluoroscopy, but the radiologist could normally only aim at its center, since the tumor expansion is not clearly visible. Thus, the obtained specimen often consisted of necrotic tissue of the tumor center, which is useless for the pathologist to specify the tumor. For biopsy, the border of the tumor is the most appropriate area to extract tissue and therefore to place the needle tip. However, since this area is not clearly visible under X-ray fluoroscopy, performing biopsies under X-ray guidance is - according to the interviewees - not very reliable.

While most radiologists explained, that they would normally use CT-guidance for needle placement, two members of the radiology group are routinely performing X-ray fluoroscopic guided needle placement, especially for spine and lung applications. Moreover, X-ray fluoroscopy guidance is recommended for the puncture of simple-shaped bones, while bones with a complex 3D structure (e.g. pelvis, shoulder, foot, etc.) are indicated to be punctured under CT-guidance, according to the interviewees. Another radiologist proposed to use the presented guiding system for training and simulation of needle placement interventions.

All interviewees explained, that in case of needle placement under X-ray fluoroscopic guidance, they would follow the *axial aiming technique*, where the C-arm detector and the X-ray source are positioned in a way, that the insertion point and the target structure are superimposed in the image (compare figure 2.6 of section 2.2).

This technique requires the distinct visibility of the target in a top- and a lateral-view, which is not always given, according to some radiologists. However, if a fluoroscopic-guided needle placement is possible from this point of view (visibility of the target from two viewing directions), this technique may have some advantages over CT-guided needle placement:

- the puncture could be performed faster than under conventional sequence-CT (but not faster than CT-fluoroscopy guided needle placement),
- it is less traumatic since it is faster than sequence-CT, and
- an X-ray C-arm is often more available than a CT scanner.

If X-ray guided needle placement with the proposed system is faster than under sequence-CT, some radiologists see the chance, that in the future more bone and lung punctures might be performed under X-ray fluoroscopy. But even if bone or lung punctures can be performed faster under X-ray fluoroscopic guidance, many radiologists that are experienced with CT-guided interventions would not easily change from their well-known procedures to a new technique, said one interviewee. Therefore, such a guiding-system must have significant advantages compared to traditional techniques.

One advantage, for example, has to be the increased precision of the puncture. Some interviewees stated, that e.g. for needle placement in the lung sometimes only the third try is successful, which is quite crucial since every attempt increases the risk for a pneumothorax. One radiologist encouraged the author to develop this *guiding system for the use in a hybrid imaging system, where an X-ray C-arm is combined with a CT scanner*. If the guiding system could be used both with the C-arm and the CT-scanner, this would be most appropriate in terms of flexibility and acceptance of the system.

Traumatology and orthopedics

The interviewed traumatologists and orthopedists explained, that there are various guiding-problems existing in traumatology. Most relevant to a guiding system would be the

- placement and distal fixation of medullary nails,
- placement and guidance of Kirschner-wires (guide-wires),
- placement of single screws, and
- guidance for drilling the fixation channels for crucial ligaments.

For these procedures usually X-ray fluoroscopy is employed to control the surgery. Therefore in this group most physicians would appreciate to have a guiding-tool in certain applications for precise placement of screws or drill holes.

Urology

Physicians in urology often have to cope with the problem, that structures of interest could not be depicted adequately with ultrasound. Especially, deep seated structures in case of stout patients can cause image resolution problems for ultrasonic imaging. In these cases X-ray fluoroscopy in combination with contrast media can be the imaging modality of choice. For urologic interventions under X-ray guidance, a guiding-system would be advantageous in order not to hold one's hands into the X-ray beam. Interventions could get simpler and more safe, according to one interviewee. Main applications for a guiding-system under X-ray fluoroscopy are seen in kidney punctures and punctures of the biliary system. In both cases the urologist first has to place a catheter or needle for initial insertion of contrast media into the bile duct or the collecting system of the kidney to make these structures visible under X-ray fluoroscopy. Afterwards, a guiding-system could help for precise percutaneous needle placement to extract e.g. gallstones or kidney stones percutaneously.

4.2.2 A Guiding-system for CT-fluoroscopy

The interviewed radiologists explained, that with conventional *sequence-CT* the puncture is performed iteratively in several steps. After the patient is moved into the CT gantry, the most appropriate image plane is chosen, which shows the best access trajectory to the target. In this image plane the insertion point, the insertion angle (angle to vertical in image) and the insertion depth of the needle is determined. Then the patient is moved out of the gantry again to perform the puncture. To support needle placement, different needle guiding systems can be used, according to the interviewees. Handheld guidance devices are not very precise but simple and widely accepted. They consist of a needle holder with an attached protractor and a level indicator, providing simple but useful support. While the puncture itself is performed stepwise outside the CT gantry, the patient has to be moved in and out of the gantry several times for control scans. This lengthy procedure is repeated till the needle has reached the target appropriately.

The needle placement technique is different, if the clinician has access to a CT scanner capable of *CT-fluoroscopy* imaging. While conventional sequence-CT scanners need at least 1 or 2 seconds for generating the image, CT-fluoroscopy is providing images with an imaging rate of up to 8 images per second. This short image acquisition time allows for manipulation of the needle directly inside the gantry, under direct image control. This approach has distinct advantages as direct visibility of the needle and tissue movements or drifting of the needle during insertion. The benefit of *direct visual feedback* faces the problem of radiation exposure and the demand for X-ray protection, if the needle is manipulated manually. A simple aid to protect the operator's hands from X-ray exposure is e.g. the use of forceps or surgical clamps to hold the needle. This, unfortunately, leads to reduced tactile feedback of the tissue during needle insertion. Furthermore, the manual manipulation of the needle inside the CT gantry would often force the physician to an inconvenient posture during the intervention, according to some interviewees.

Most of the interviewed radiologists stated to see a need for a guiding system for CT-fluoroscopy, to 'replace' the physicians hand holding the needle inside the gantry. But its usage should be simple and fast. Precise preoperative planning and determination of the access trajectory to the target is important, especially in the region of high risk organs, like large vessels or nerves. Guidance of the needle is desirable especially in case of double-oblique needle orientation[18] or tilted CT-gantry - circumstances that are normally avoided because of unsatisfactory means of guidance. On the other hand, just *the use of such a guiding system could make these punctures more reliable, easier to perform and therefore more common in the future.*

While one half of the interviewed radiologists emphasized the importance of tactile feedback during manual needle insertion, the other half could imagine even to use a joystick for remotely controlled manipulation and insertion.

Advantages of a guiding system should be shorter biopsy duration (or more specimen in the same time), higher precision and high system-safety. A remotely controlled system would be desirable with regard to dose reduction for the physician. One radiologist even suggested to 'shoot' the needle through the tissue towards the target to minimize the duration for needle insertion in order to avoid patient movement during the procedure (patient has 'no time' to wince or move). However, the physician must have the complete control over both, the system and the patient, all the time during the procedure. If the needle is drifting away from the desired trajectory during insertion, the physician must be able to stop the insertion and correct the path.

The interviewees explained, that *laser guiding systems* are increasingly used to support in percutaneous needle placement (compare section 2.3.1). After the target and skin entry point are defined in the CT-scans, the system is computing the desired geometry parameters which are needed to adjust the laser. Then the needle is placed in the laser beam, which is providing precise orientation during manual needle insertion. Laser-guided needle placement is always performed outside the CT-gantry.

4.2.3 Further Requirements Stated by the Interviewees

All interviewees emphasized that a simple operation of the guiding-system is essential. The operating instructions should not be longer than two A4 pages. The sterilization of parts, being in direct contact with the patient, is considered to be very important. All other parts, like the passive arm or cables should be covered by surgical drape before every operation. An input device during the intervention could be e.g. a sterile mouse. The length of the guiding cannula should not be longer than 5 cm.

---

[18] Double-oblique with regard to the image plane, which means tilted
      to the image vertical and tilted out of the image plane.

Some radiologists explained, that needle placement in or near the thorax is difficult due to patient breathing during the puncture. Patient breathing can move the target tissue or the needle out of the scan plane where they are no longer visible in the CT image. These moving tissues may cause problems if the guiding system tries to keep the needle in a fixed orientation. If the needle alignment process via guiding system would not take longer than 20 to 30 seconds, it could be performed during a single breathhold. Additionally, the system has to be capable of unlocking the needle after the desired needle location has been reached.

Bringing such a guiding-system on the market, the system's design (look-and-feel) is important as well, according to one interviewee. Including hardware and a software package, a system price of about US$ 25.000 would be acceptable.

4.3 Summary of Clinical Requirements

All these information about potential applications of a robotic needle guiding system are leading to certain clinical requirements for the automatic needle alignment procedure as well as for the guiding-system itself. Table 4.3 summarizes the basic clinical requirements for the needle placement procedure. Most important seems to be the aspect of (a) increased precision, (b) better control of the procedure, and (c) dose reduction for the physician.

One can define many more requirements from all proposed guiding applications, than listed in this section. A guiding system specialized e.g. for placement of a bone-drill has obviously to meet other technical requirements than a system optimized for placement of a simple puncture needle, even if the image guided alignment approach may be identical for both applications. Since this thesis sets focus on automatic image guided needle placement, all requirements summarized in table 4.3 refer to percutaneous needle biopsies and punctures.

summary of clinical requirements
• increased precision of the puncture compared to manual needle placement
• punctures - too complicated or dangerous before - should get feasible
• shorter biopsy duration than manual techniques
• less complications compared to manual techniques
• does reduction for the physician (remote or automatic control of the needle)
• procedure should be capable to integrate C-arm- and CT-imaging
• sterilization of the end-effector

Table 4.3: Summary of clinical requirements for the automatic needle placement procedure, acquired from the physician interviews [7].

Chapter 5
Development of a Needle Guiding Robot

Broad acceptance and success of a robotic needle-guiding system on the health care marked strongly depends on its potential to improve patient outcome in a cost-effective manner. Therefore, it is firstly essential that the system satisfies the physicians' needs, and secondly the usage of the system must fit appropriately into the conventional workflow the physician is used to. Due to a faster and improved workflow, or the ability to perform needle placement procedures in high risk areas, the system may have the potential to reduced in-hospital costs as well.

The following chapter describes the development and realization of a miniaturized robotic system for automatic image-guided needle placement for X-ray fluoroscopy and CT imaging. First, the clinical requirements, which have been presented in chapter 4, will be translated into *technical requirements* the robotic system has to meet. The second section sets focus on the *hardware design*, like the special manipulator cinematic, the integrated sensors and cabling. This is followed by a mathematical description of the *robot's kinematic*, which is needed for the image-based control design (visual servoing) presented in the chapter 6.

5.1 The Technical Requirements for the Needle Guiding Robot

The physician interviews presented in chapter 4 have led to certain *clinical requirements* summarized in table 4.3. These clinical requirements however are taken into account for the definition of specific *technical requirements* for the needle guiding robot which are presented in table 5.1.

---

technical requirements for the needle guiding robot

- rigid and precise design of the guiding-system
- active needle motion envelope should be a cone with at least ±45°
- need of as less space as possible for the robot to move the needle
- simple disconnection of the needle from the system after needle placement
- sterilization of all parts that are in direct contact with the patient
- maximum length of the guiding cannula should be less than 5cm
- manual and remotely controlled needle insertion should be possible
- need to keep the guiding cannula precisely in place during needle insertion
- software tool for pre-interventional planning and controlling of the puncture
- system must provide physician complete control over the procedure all the time
- suitable for usage with a C-arm fluoroscope or within a CT scanner

---

Table 5.1: Summary of technical requirements for the robotic guiding-system, resulting from the clinical requirements described in section 4.3 and the physician interviews [7].

Beside the fundamental needs like a precise and rigid design of the manipulator links, requirements concerning the robot's needle holder (guiding cannula) are essential, such as the sterilization of the needle-holder and the possibility to use different needle diameters. Furthermore, the length of the guiding cannula should not exceed 5 cm in order to limit the required length for the employed puncture needles. The required tilt-angles of the needle around the insertion point (motion envelope), shall be at least ±45° (compare figure 5.9). For a wide range of application and different regions of insertion, it is advantageous that the robot requires as less space for moving the needle around the insertion point as possible. Furthermore, as an alternative to manual needle insertion, the robot shall provide a special needle drive mechanism to allow for remotely controlled needle insertion.

Because of safety reasons it is essential that the physician has the absolute control over the system during the automatic alignment process and the remotely controlled needle insertion. For example, the immediate stop of the robot must be possible in case of failure. Therefore, a special software tool for planning and controlling the whole procedure is necessary. For accurate needle insertion, regardless whether manually or remotely controlled, it is essential that the robot keeps the aligned guiding cannula precisely in place. The complete system must be suitable for the usage under the guidance of X-ray fluoroscopy as well as CT imaging, which is leading to the need for a radiolucent distal part of the robot near the end-effector.

All these technical requirements have been taken into account during the development of the needle guiding robot, which is described in detail in the following sections.

5.2   Development of the Robot's Design

The goal was to develop an optimized needle guiding system for automatic or remotely controlled percutaneous needle placement, appropriate for the usage together with an X-ray C-arm or within a CT scanner. Furthermore, the robot has to meet all technical requirements which have been defined above.

As discussed in section 3.1 the work envelope[19] of a manipulator is the 3D space reachable for the robot to place its end-effector for performing the programmed task. The shape of the work envelope basically depends on the kinematic structure and the geometry of the robot. In case of the presented needle placement application, the task to be performed by the robot is to align a needle in space with a certain target. This has to be accomplished by rotating the needle around a fixed rotation point in space, i.e. the needle insertion point on the patient's skin. Therefore, the *work envelope* of the robot is reduced to a *conical space* wherein the needle can be rotated by the robot.

Figure 5.1 shows three different kinematic structures for a manipulator appropriate to rotate a needle in 3D around a fixed rotation point (remote center of motion). The first example (a) represents an arc-based structure. The guiding cannula is fixed to a slider which is carried on a circular arc that, in turn, is mounted on a rotatable ring. The remote center of motion is geometrically defined by the physical design (center of arc) and requires only two degrees of freedom (*dof*) for placing the needle. Example (b) illustrates a typical serial-link kinematic, similar to that commonly used in many industrial robots. The pose of the guiding cannula is the result of the collective effect of rotation of all four manipulator joints. Therefore, this serial-link kinematic demands high controller performance since each needle rotation requires a complicated interaction of all joints. In this example the rotation of the needle around a remote center of motion requires four *dof*. The manipulator principle presented in example (c) is based on a parallel kinematic. This special geometry effects a simultaneous, parallel tilting of all vertical links, including the guiding cannula. As in example (a) the needle motion is restricted to a rotation around a fixed center, wherefore only two *dof* are required.

---

[19] In literature the work envelope is often referred to as the 'workspace' of a robot. However, in this thesis the workspace of the presented system is equivalent to the reachable space of the needle guide by moving manually the passive arm (compare 5.9). In contrast to this, the work envelope is the small, conical space wherein the robot can rotate the needle.

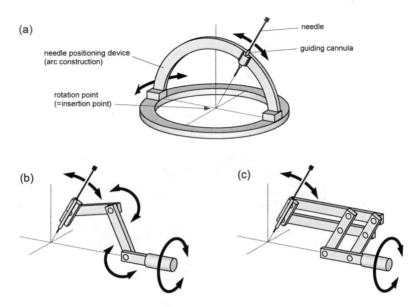

Figure 5.1:   Principles of a needle guiding system with different kinematic structures. All of them are appropriate for rotating a needle in 3D around a fixed rotation point (insertion point of the needle on the patient's skin): (a) arc-based structure, (b) serial-link structure, (c) parallel kinematic.

In common robot applications a serial-link manipulator is advantageous in terms of a large workspace or high flexibility in positioning the end-effector. However, in the presented needle placement application the task which has to be performed is to rotate a needle in 3D around a remote center of motion. This is a very restricted and simple motion of the end-effector (guiding cannula) which requires only two *dof* for orienting the needle. Therefore, the use of a serial-link manipulator as shown in example (b) is needless. Additionally, it requires a sophisticated controller and, according to its *dof*, four joint actuators. In contrast, example (a) and (c) require only two actuators for orienting the needle. Furthermore, the kinematic structure of (a) and (b) constrains the robot motions to a well defined limited area of operation. This is advantageous in the event of failure, since a manipulator with many *dof* and designed to have a large envelope of motions is intrinsically less safe than a *special purpose manipulator* whose motions and forces were designed specifically for the task [42]. Comparing the kinematic structures of example (a) and (c), it is obvious that (c) can be realized in a more compact design than (a). Additionally, the parallel kinematic provides more space near the guiding cannula while permitting a large angle range for rotating the needle. The needle guiding system must be as radiolucent as possible close to the guiding cannula in order not to obscure the patients anatomy in the X-ray images. This is recommended especially when working with an X-ray fluoroscope. Therefore, motors or other metallic parts have to be located away from the guiding cannula. This requirement is supposed to be achieved more easily in example (c) than in (a).

These considerations led to the decision to go for a needle guiding robot with a *parallel kinematic* as shown in figure 5.1-c. There are different examples existing in the medical field where manipulators with a parallel kinematic are employed. Especially in laparoscopic applications the desired task is very similar: a straight endoscopic instrument has to be rotated while been placed through a small incision in the abdominal wall. An example of such a laparoscopic manipulator with a parallel kinematic is the passive TISKA Endoarm (KARL STORZ Endoskope GmbH, Tuttlingen, Germany) originally developed by the Karlsruhe Research Center, Germany [137]. Figure 5.2 shows the system in an experimental setup. A further example of a parallel kinematic manipulator is the LARS robot (see figure 3.8), which already has been presented in section 3.2 [158].

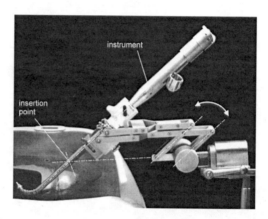

Figure 5.2: TISKA Endoarm (KARL STORZ Endoskope GmbH, Tuttlingen, Germany). This passive positioning arm can rotate an endoscopic instrument placed through a small incision in the abdominal wall (insertion point). It is required that the remote center of motion of the parallel kinematic is superimposed with the insertion point. In the desired position the system can be fixed with electromagnetic brakes.

5.2.1 Structure of the Robot

Following the considerations above, a compact surgical robotic system has been developed comprising a parallel kinematic for needle rotation. A draft of this novel micro-robot is shown in figure 5.3.

Its very small and compact design (length 365 mm, weight 590 g) with a radiolucent distal part is optimized for precise needle manipulation inside a CT-gantry or together with an X-ray C-arm. The construction consists of a very compact parallel linkage structure that allows needle rotation around two perpendicular axes. These two rotational degrees of freedom allow to rotate the needle in 3D around the remote center of motion, which is given by the geometry of the design. For automatic needle placement the operator has to position the robot on the patient so that the remote center of motion is identical with the needle insertion point on the patient. To support this, the guiding cannula is designed in a way that its tip is equivalent to the remote center of motion. Therefore, the operator just has to place the tip of guiding cannula at the desired insertion point on the patient's skin.

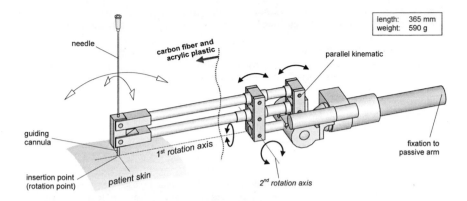

Figure 5.3: The robot shows two perpendicular rotational axes that allow 3D needle rotation around a fixed rotation point. The parallel kinematic enables a very compact design appropriate for usage inside a CT scanner. The complete system is very small and lightweight designed with provides enough space between robot and the patient, even for strongly tilted needles. All cables are integrated inside the manipulator.

The parallel kinematic allows the easy separation between a radiolucent distal part, where only carbon fiber and acrylic plastic is used (except the metallic guiding cannula), and a very rigid metallic structure containing all servo drives, encoders and limit switches (compare figure 5.3). Apart from CT-guided interventions, this radiolucent distal part of the robot makes X-ray imaging (fluoroscopy) possible, where a needle guiding tool should not obscure the patient's anatomy in the X-ray image. All electric cables for motors, encoders and limit switches are integrated inside the robot housing. Additionally, a special radiolucent needle transmission is integrated at the guiding cannula, which allows for remotely controlled needle insertion (compare section 5.2.2).

The structure of the robot basically consists of two components, a *static part* and a *rotating part* (see figure 5.4). The rotating part contains the parallel kinematic and the end-effector, and is rotatable connected with the static part by the main shaft. The static part is connected to

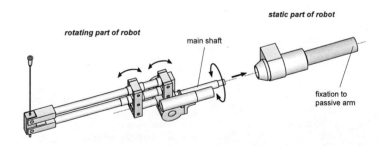

Figure 5.4: The structure of the robot basically consists of two components, a *static part* and a *rotating part*. The rotating part contains the parallel kinematic and the end-effector, and is rotatable connected with the static part by the main shaft.

a passive arm and integrates the DC motor, gear, axis encoder and micro limit switches in order to rotate and control the main shaft. An axial section through the robot is shown in figure 5.5.

The proximal end of the robot can be fixed to the passive arm and integrates a 28-pin female plug connector for the electric contact with the arm. Adjacent to the plug connector there is a DC servo motor (drive 1) located with integrated motor encoder and a planetary gear (2.6 W, gear ratio 245:1, FAULHABER GmbH, Schoenaich, Germany). Drive 1 is connected with the main shaft via coupling. The main shaft is precisely bedded with two angular ball bearings, which are preloaded with an adjusting nut. For precise measurement of the rotatory orientation of the main shaft a small high resolution encoder (SPTS, MEGATRON Elektronic, Putzbrunn, Germany) is mounted directly on the main shaft between the angular ball bearings. This optical, incremental encoder provides a resolution of $\pm 0.04°$. Additionally, two micro switches were integrated in the housing of the static part of the robot (ultraminiature switch, $8.2 \times 6.2 \times 2.7$ mm, Cherry Electrical Products, Pleasant Prairie, USA). They serve as limit switches and get pushed by a pin on the main shaft in case of a shaft rotation of $\pm 60°$. The micro switches are automatically checked by the employed DC motor encoder board (compare section 6.6.1). Due to the gear friction and the high gear ratio the rotating part of the robot keeps in place without an additional break even under the application of external forces.

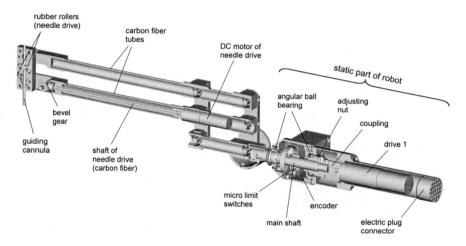

Figure 5.5: Axial section through the robot. The parallel structure with its end-effector and integrated needle drive is mounted on the main shaft. The main shaft is bedded with two angular ball bearings and is rotated by a DC servo motor inside the static part of the robot.

The rotating part of the robot is mounted on the main shaft. It consists of the parallel kinematic structure and the end-effector with the needle drive. Figure 5.6 shows the power transmission for the parallel kinematic structure. It is actuated by a small DC servo motor (drive 2) with integrated motor encoder and a planetary gear (2.6 W, gear ratio 14:1, FAULHABER GmbH, Schoenaich, Germany). The servo motor drives a small worm gear (gear ratio 30:1). The worm wheel is mounted on a shaft and rotates two vertical rods. This causes the tilting of the whole parallel structure including the end-effector with the guiding cannula. Due to the self-locking capability of the worm gear there is no additional break required.

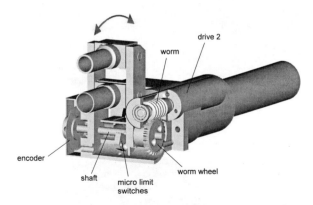

Figure 5.6: Power transmission for tilting of the parallel structure. It is actuated by a small DC servo motor (drive 2) and a worm gear. The worm wheel is mounted on a shaft and rotates two vertical rods which causes the tilting of the parallel structure.

The main shaft is bedded with two micro ball bearings. Again, a small high resolution encoder (SPTS, MEGATRON Elektronic, Putzbrunn, Germany) directly mounted on the shaft allows to determine precisely the tilt angle of the parallel structure. The same type of micro limit switches as used for the main shaft (see above) are employed for the worm gear in order to restrict the maximum tilt angle of the parallel structure to ±50°.

All wires for the three DC motors, the integrated motor encoders, the axis encoders, and the limit switches run inside the robot housing. Between the rotating and the static part of the robot, the wires run through the main shaft which is partly hollow. The integration of all wires inside the manipulator allows easy cleaning and coverage of the robot, e.g. with surgical drape, before a sterile needle placement procedure.

5.2.2 The Needle Drive

The end-effector has integrated a very compact radiolucent needle drive which allows for remotely controlled needle insertion. It is powered by a micro DC motor (0.65W, gear ratio 64:1, FAULHABER GmbH, Schoenaich, Germany) integrated in the parallel structure outside the radiolucent area (compare figure 5.5). The DC motor drives a thin carbon fiber shaft. A following bevel gear and a belt transmission transfers the rotation into the end-effector of the robot. Figure 5.7 shows a cross-section of the needle drive. Four micro gearwheels transmit the applied rotation to special rubber rollers which are driving the needle. The guiding cannula is split in two parts, an upper and lower part, with the two rubber rollers in-between (compare figure 5.5 and 5.7). The needle which is placed in the guiding cannula is automatically grasped between both rubber rollers that are working as a kind of 'friction gear'. Since the needle is exclusively driven by friction between the rubber rollers and the needle, it can apply only moderate axial forces during needle insertion. However, a special feature of the robot's needle drive is that it grasps the distal end of the needle, not the needle head. This significantly reduces the unsupported length of the needle and thus minimizes the lateral flexure during injection and increases the accuracy of the puncture. A similar friction transmission for a needle drive has been developed and evaluated by Stoianovici et al. for the PAKY robot [152].

Tilting the parallel structure around the 2nd rotation axis causes a small, unintended rotation of the transmission belt. Although this effect is only small it gets automatically corrected by the robot control.

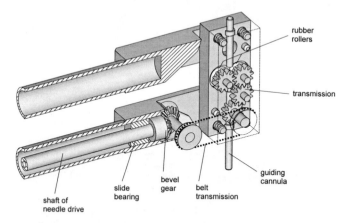

Figure 5.7:   Cross-section of the needle drive integrated in the end-effector. The rotation is trans-
mitted by a bevel gear, a belt transmission, and several gearwheels to two rubber rollers. These rubber rollers grasp the needle and drive it due to friction.

5.2.3 The Passive Manipulator Arm

The work envelope of the robot, in which the needle can be rotated, is very compact. Translational motions cannot be performed by the manipulator. Therefore, an additional stage or arm is required for holding and positioning the robot in space. Figure 5.8 shows the passive arm which has been developed in order to position manually the robot at the desired location. To-

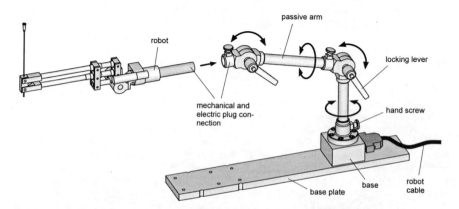

Figure 5.8:   The passive manipulator arm for holding and positioning the robot. The arm can be
manually locked in the desired position by hand screws and locking levers. All cables
are integrated within the passive arm.

gether with this positioning arm the robot works more as an 'end-effector' for needle rotation, while the passive arm enables the operator to position the needle guide within the 'workspace' of the arm. The passive arm provides four degrees-of-freedom and can be manually locked in the desired position by hand screws and locking levers. The robot can be easily attached and detached to the arm while the electric contact between arm and robot is obtained by a special electric plug connection. The passive arm allows to position the robot within a sphere with a radius of about 600 millimeters.

All wires are integrated within the links and joints of the passive arm and run down to the base where the connector for the robot cable is located. This concept of integrated wires, which already has been implemented for the needle guiding robot, allows easy handling and is advantageous in terms of cleaning or covering with surgical drape.

5.3 Mathematical Description of the Robot's Kinematic

The following sections present the mathematical description of the robot's kinematic. Purpose is to describe the needle pose in the robot coordinate frame my means of mathematical equations depending on the motor motions. This is of basic interest for the development of the robot control software, which is presented in section 6.6.2.

The actual needle pose is described in a *direct kinematics* approach (compare section 3.1.2) by several consecutive coordinate transformations. While in common robot applications the end-effector pose has to be described by homogenous transformation matrices, representing rotational and translational motions, the presented needle placement application can be simply described by rotation matrices, following the Euler's theorem of rotation [143].

5.3.1 Representation of Needle Orientation with Euler Angles ψ, θ, ϕ

The manipulator has to tilt the needle around a fixed-point in space, (*remote center of motion*), which is identical with the needle insertion point through the patient's skin (see figure 5.9). The needle axis is described in the end-effector coordinate system N. The initial needle position is equivalent to the z-axis of a cartesian coordinate frame. Its tip is located at the base of this coordinate system and is corresponding to the center of rotation. For medical applications a maximum needle tilt range of at least ±45° is required. Therefore, the robot's kinematic is designed for a needle work envelope with a tilt range of ±50°, corresponding to a

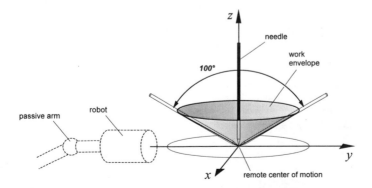

Figure 5.9: The work envelope of the needle. The initial needle position is identical with the z-axis while the needle tip is representing the origin of the cartesian coordinate frame.

circular cone with a vertex angle of 100° and the z-axis corresponding to the axis of symmetry.

The mathematical representation of the needle orientation is derived from the Euler's theorem of rotation using the Euler angles ψ, θ, ϕ (see figure 5.10) [143]. The advantage of this representation is that a planar needle motion can be easily described by a plane $\pi(\psi, \theta)$ and the angle ϕ. Usually the Euler notation describes a zxz-rotation, which represents three consecutive rotations around the z-axis, the x-axis and the obtained z-axis again [107]. However, for the presented application it is more appropriate to use a zxy-rotation to describe the needle orientation, with the needle axis equivalent to the z-axis of the frame.

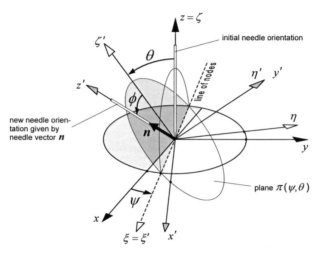

Figure 5.10:   Geometrical description of the needle orientation using Euler angles ψ, θ, ϕ, which are representing a zxy-rotation. Euler coordinate transformation: (1) rotation of initial coordinate frame (x,y,z) with ψ around z-axis; (2) rotation of derived frame (ξ, η, ζ) with θ around ξ-axis (*line of nodes*); (3) final rotation of derived frame (ξ', η', ζ') with ϕ around axis η', yielding the final orientation (x',y',z').

The needle orientation is given by a normalized vector n which is identical with the z-axis in its initial pose. The rotational transformation of this initial frame (x,y,z) into any arbitrary frame, where the z-axis lies inside the work envelope, will be described with three consecutive rotations. First the original frame (x,y,z) is rotated counterclockwise around the z-axis with angle ψ, yielding a new frame (ξ, η, ζ). This frame (ξ, η, ζ) is now rotated around axis ξ (*line of nodes*) with angle θ, which is leading to a second transitional frame (ξ', η', ζ'). The desired coordinate frame (x',y',z') is obtained after a third rotation of frame (ξ', η', ζ') with angle ϕ around axis η'. These three consecutive rotations can be derived as a multiplication of three rotational matrices. The first rotation is given by

$$[x, y, z]^T = R_z [\xi, \eta, \zeta]^T \quad \text{with } R_z = \begin{bmatrix} \cos\psi & -\sin\psi & 0 \\ \sin\psi & \cos\psi & 0 \\ 0 & 0 & 1 \end{bmatrix} \quad (5.1)$$

and that the whole rotational transformation is represented by

$$[x, y, z]^T = \mathbf{R}_z \mathbf{R}_\xi \mathbf{R}_{\eta'} [x', y', z']^T. \tag{5.2}$$

With each single rotation matrix

$$\mathbf{R}_z = \begin{bmatrix} \cos\psi & -\sin\psi & 0 \\ \sin\psi & \cos\psi & 0 \\ 0 & 0 & 1 \end{bmatrix} \quad \mathbf{R}_\xi = \begin{bmatrix} 1 & 0 & 0 \\ 0 & \cos\theta & -\sin\theta \\ 0 & \sin\theta & \cos\theta \end{bmatrix} \quad \mathbf{R}_{\eta'} = \begin{bmatrix} \cos\phi & 0 & \sin\phi \\ 0 & 1 & 0 \\ -\sin\phi & 0 & \cos\phi \end{bmatrix} \tag{5.3}$$

one derives the complete Euler rotation, representing a needle orientation as described above:

$$\mathbf{R}(\psi,\theta,\phi) = \mathbf{R}_z \mathbf{R}_\xi \mathbf{R}_{\eta'} = \begin{bmatrix} \cos\psi\cos\phi - \sin\psi\sin\theta\sin\phi & -\sin\psi\cos\theta & \cos\psi\sin\phi + \sin\psi\sin\theta\cos\phi \\ \sin\psi\cos\phi + \cos\psi\sin\theta\sin\phi & \cos\psi\cos\theta & \sin\psi\sin\phi - \cos\psi\sin\theta\cos\phi \\ -\cos\theta\sin\phi & \sin\theta & \cos\theta\cos\phi \end{bmatrix} \tag{5.4}$$

The columns of this matrix are equivalent to the base vectors (x',y',z') of the new coordinate frame, described in (x,y,z) coordinates. Since the needle orientation $\mathbf{n}$ is the z'-axis by definition, the needle vector is obtained from the last column of the Euler rotation matrix:

$$\mathbf{n}(\psi,\theta,\phi) = \mathbf{R}(\psi,\theta,\phi) \begin{pmatrix} 0 \\ 0 \\ 1 \end{pmatrix} = \begin{bmatrix} \cos\psi\sin\phi + \sin\psi\sin\theta\cos\phi \\ \sin\psi\sin\phi - \cos\psi\sin\theta\cos\phi \\ \cos\theta\cos\phi \end{bmatrix} \tag{5.5}$$

5.3.2 Representation of Needle Orientation with Motor Angles α, β

The needle guiding robot is provided with two motors in order to rotate the needle around the remote center of motion, as shown in figure 5.11. Drive 1 is fixed to the passive arm. The rotation axis is equivalent to the y-axis of the initial coordinate frame (x,y,z). While the needle

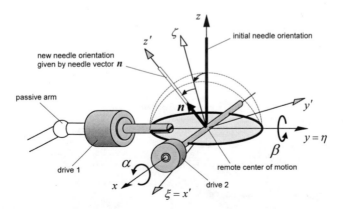

Figure 5.11: The needle gets rotated around the remote center of motion by two drives which are arranged perpendicular to each other. Thus, the needle motion directly follows the Euler's theorem of rotation.

is fixed to the second motor's shaft (x-axis), a turn of the first motor's shaft causes a rotation of both motor 2 and the needle around the y-axis. Due to this combination of two perpendicular rotational axes y and x, the needle can be positioned in any orientation in space. While commonly three rotational degrees of freedom are necessary to perform an arbitrary rotation of a rigid body, here the third rotational degree of freedom would be needed to turn the needle around its own axis z, which is needless in this application. These two rotations around the y- and ξ-axis can be described in matrix form as

$$\boldsymbol{R}_y = \begin{bmatrix} \cos\beta & 0 & \sin\beta \\ 0 & 1 & 0 \\ -\sin\beta & 0 & \cos\beta \end{bmatrix} \quad \boldsymbol{R}_\xi = \begin{bmatrix} 1 & 0 & 0 \\ 0 & \cos\alpha & -\sin\alpha \\ 0 & \sin\alpha & \cos\alpha \end{bmatrix} \tag{5.6}$$

The entire coordinate transformation is achieved by multiplication of $\boldsymbol{R}_y$ and $\boldsymbol{R}_\xi$

$$[x,y,z]^T = \boldsymbol{R}_y\boldsymbol{R}_\xi[x',y',z']^T = \begin{bmatrix} \cos\beta & \sin\beta\sin\alpha & \sin\beta\cos\alpha \\ 0 & \cos\alpha & -\sin\alpha \\ -\sin\beta & \cos\beta\sin\alpha & \cos\beta\cos\alpha \end{bmatrix}[x',y',z']^T. \tag{5.7}$$

Again, the columns of the rotation matrix $\boldsymbol{R}_y\boldsymbol{R}_\xi$ are representing the base vectors (x',y',z') of the new coordinate frame with z' as new needle axis. Therefore, the needle vector $\boldsymbol{n}$ is obtained from the last column of this rotation matrix

$$\boldsymbol{n}(\alpha,\beta) = \begin{bmatrix} \sin\beta\cos\alpha \\ -\sin\alpha \\ \cos\beta\cos\alpha \end{bmatrix}. \tag{5.8}$$

5.3.3 Computation of Motor Angles α, β out of the Euler Angles ψ, θ, ϕ

After equating (5.5) with (5.8) the motor angles α and β can be easily determined out of the Euler angles ψ, θ, ϕ:

$$\boldsymbol{n}(\psi,\theta,\phi) = \boldsymbol{n}(\alpha,\beta) \tag{5.9}$$

$$\sin(\alpha) = -\sin(\psi)\sin(\phi) + \cos(\psi)\sin(\theta)\cos(\phi) \tag{5.10}$$

$$\tan(\beta) = \frac{\cos\psi\sin\phi + \sin\psi\sin\theta\cos\phi}{\cos\theta\cos\phi} \tag{5.11}$$

$$\alpha(\psi,\theta,\phi) = \arcsin(-\sin(\psi)\sin(\phi) + \cos(\psi)\sin(\theta)\cos(\phi)) \tag{5.12}$$

$$\beta(\psi,\theta,\phi) = \arctan\left(\tan\phi\frac{\cos\psi}{\cos\theta} + \sin\psi\tan\theta\right) \tag{5.13}$$

The angles α and β are of significant interest for programming the motor control. After defining mathematically the new needle orientation $n(\psi,\theta,\phi)$ in Euler angle representation, the required motor angles $\alpha(\psi,\theta,\phi)$ and $\beta(\psi,\theta,\phi)$ can directly be derived by equation (5.12) and (5.13).

5.3.4 Computation of ψ_{new} and θ_{new} out of Two Needle Orientations

The needle alignment algorithm using X-ray fluoroscopy imaging, which will be presented in section 6.3.1, requires the computation of a plane π_{new} spanned by two needle vectors n_1 and n_2 (see figure 5.12).

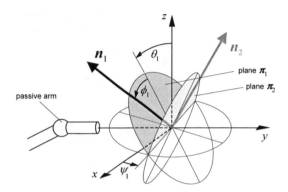

Figure 5.12:   Two needle orientations represented by their needle vectors n_1 and n_2 in Euler angles.

The needle vectors $n_1(\psi_1,\theta_1,\phi_1)$ and $n_2(\psi_2,\theta_2,\phi_2)$ have been derived by needle rotation within the two planes π_1 and π_2. The new plane π_{new} is defined by ψ_{new} and θ_{new}, which can be derived by the normal vector n spanned by n_1 and n_2.

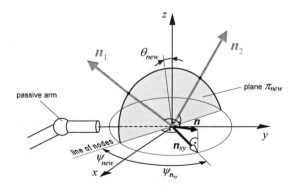

Figure 5.13:   The plain π_{new} is spanned by the needle vectors n_1 and n_2.

The normal vector $\boldsymbol{n}$ of plane π_{new} spanned by $\boldsymbol{n}_1$ and $\boldsymbol{n}_2$ is defined as

$$\boldsymbol{n} = \boldsymbol{n}_1 \times \boldsymbol{n}_2 = \left(n_x, n_y, n_z\right)^T. \tag{5.14}$$

The normal projection of $\boldsymbol{n}$ onto the xy-plane is given by

$$\boldsymbol{n}_{xy} = \left(n_x, n_y, 0\right)^T. \tag{5.15}$$

Notice that the direction of normal vector $\boldsymbol{n}$ is inconsistent, i.e. by commutation of both needle vectors $\boldsymbol{n}_1$ and $\boldsymbol{n}_2$ the sign of $\boldsymbol{n}$ is changing. For the computation of angle θ_{new} it is necessary to have a normal vector independent of the order of both needle vectors. Therefore the sign of $\boldsymbol{n}$ is switched if n_y is negative (compare (5.17)). Thus $\boldsymbol{n}$ lies always in the positive half-space of the y-axis.

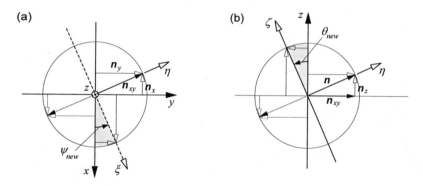

Figure 5.14: Computation of the desired rotation angles ψ_{new} and θ_{new}.

The desired rotation angle ψ_{new} is the angle between the y-axis and vector $\boldsymbol{n}_{xy}$ (compare figure 5.14-a) and can be computed as

$$\psi_{new} = \arctan\left(\frac{-\boldsymbol{n}_x}{\boldsymbol{n}_y}\right). \tag{5.16}$$

The negative sign leads to the mathematically correct rotation around the z-axis. Angle θ_{new} between the two vectors $\boldsymbol{n}$ and $\boldsymbol{n}_{xy}$ (see 5.14-b) can be easily obtained by equation

$$\theta_{new} = \arctan\left(\frac{\boldsymbol{n}_y}{|\boldsymbol{n}_y|} \cdot \frac{\boldsymbol{n}_z}{|\boldsymbol{n}_{xy}|}\right), \tag{5.17}$$

while $\dfrac{\boldsymbol{n}_y}{|\boldsymbol{n}_y|}$ is representing the sign for the eventual sign switch to achieve the mathematically

correct rotation direction around the ξ-axis.

5.4 Mechanical Stiffness of the Robot

The stiffness of the robot's mechanical design is important for the precise placement of the needle. It has been tested in lateral and vertical direction. While the proximal end of the robot is fixed with the vise of a milling machine, defined test forces F in lateral and vertical direction are applied to the distal, free end (see figure 5.15). The robot's deflection δ is measured with a dial gage and is used to determine the elastic characteristic of the whole construction in lateral and vertical direction.

Figure 5.15: Stiffness test of the robot in lateral direction. The proximal end of the robot is fixed with the vise of a milling machine, while defined test forces F are applied to the distal free end. The deflection δ of the robot is measured with a dial gage.

The results of these stiffness tests are shown in figure 5.16. Since the elastic characteristic of the robot is supposed to be linear, the measured points are fitted with a line.

The stiffness of the robot can be described by its *spring constant* d_{robot}, which is equivalent to the reciprocal of the gradient m of the fitted lines:

$$d_{robot} = \frac{F}{\delta}, \qquad \text{with} \quad \begin{aligned} d_{\text{lateral}} &= 5.15 \ N/mm \\ d_{\text{vertical}} &= 7.40 \ N/mm \end{aligned} \qquad (5.18)$$

To illustrate the values of d_{lateral} and d_{vertical}, they are compared with a metal tube with the same length as the robot and the same spring constant for lateral bending.

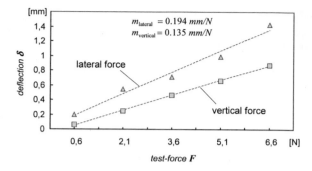

Figure 5.16:   Measured deflections δ of the robot in lateral and vertical direction after applying defined test forces F. The elastic characteristic of the robot should follow a
linear function. Therefore the measured deflections are approximated with a line,
of which the gradient m is characteristic for the robot's stiffness.

Using the maximum deflection δ_{max} of a one-sided fixed tube (inner/outside diameter D_i, D_a)
under a lateral force F

$$\delta_{max,\,tube} = \frac{FL^3}{3EI}, \qquad \text{with} \quad \begin{aligned} L &= 365 \; mm \\ E &= 210\,000 \; N/mm^2 \\ I &= \pi/64\,(D_a^{\,4}-D_i^{\,4}) \end{aligned} \qquad (5.19)$$

its spring constant can be computed as follows:

$$d_{tube} = \frac{F}{\delta_{max,\,tube}} = \frac{3EI}{L^3} = d_{robot}. \qquad (5.20)$$

After setting $d_{tube}=d_{robot}$ and solving for D_a leads to $D_{a,\,lateral}=15.9$mm and $D_{a,\,vertical}=18.4$mm
(with $D_a-D_i = 0.5$), i.e. the robot's stiffness is comparable to that of a tube with a diameter of
about 17mm (inner diameter 16.5mm). Since the robot is designed only for manipulation of
needles, this stiffness should be more than sufficient.

5.5   Mechanical Accuracy of the Robot

Naturally, the robot can position the needle around the remote center of motion only with a
limited accuracy. Reasons of inaccuracies involve gear backlash, sensor precision (resolution
of motor- and axis-encoders), servo precision of the motor controller, elastic compliance and
vibrations (compare figure 5.17).

Gear backlash of the two main drives (1st and 2nd rotation axis) is about ±1°. The internal
motor encoder provides a resolution of 512 pulses per revolution, leading to an encoder inac-
curacy of ±0.35°. However, motor encoder resolution does not really affect the needle pose,
since these inaccuracies are reduced by the motor gear for more than 240 times. To improve
the robot's mechanical accuracy in orientating the needle, two small high resolution encoders
are mounted directly on the 1st and 2nd rotation axis (SPTS, MEGATRON Elektronic, Putzbrunn, Germany). These optical, incremental encoders provide two TTL output channels with

a resolution of 1024 pulses per revolution. Together with the pulse acquisition PC plug-in board (compare section 6.6.1), which accomplishes a quadruplication of the increments, a sensor resolution in needle rotation of ±0.04° is achieved.

In the presented needle placement application, the needle is initially moved stepwise during intermittent imaging. This process is exclusively controlled by the image features (needle pose and location of the target), which are extracted out of the captured X-ray or CT images. At this time, the needle alignment accuracy basically depends on precise image analysis, whereas the *absolute accuracy* of the manipulator is of lower interest. However, as soon as there is needle alignment achieved in the images, the actual needle pose must be determined by the robot. This has to be done very precisely, since these angles are used for further computations of the desired 3D needle pose. However, accurate determination of the actual needle orientation can be precisely done by the axis encoders with an accuracy of ±0.04°.

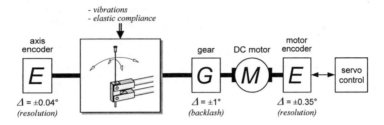

Figure 5.17: Several factors have impact on the mechanical accuracy of the robot: (i) the transmission chain: gear backlash, motor encoder resolution, axis encoder resolution; (ii) static and dynamic disturbance: vibrations, elastic compliance.

After calculation of the final 3D needle orientation, the robot has to move and adjust the needle in this desired pose with high precision. The motor angles are precisely executed by the servo control. But due to gear backlash, the result would be only of moderate accuracy. Again, the precise axis encoders are employed to improve the needle positioning accuracy. A control loop is implemented in the robot control software which analyzes the axis encoders and drives the DC motors via servo control commands as long as the axis encoders indicate accurate achievement of the desired needle orientation. Accordingly, accurate determination and execution of a needle pose is possible, despite gear backlash, due to the use of precise axis encoders with a precision of ±0.04°.

Further factors which have impact on the overall accuracy of the alignment process are image distortions and the precision in image feature extraction (needle pose, target location). However, inaccuracies in needle placement caused by *target motions*, e.g. caused by *patient breathing*, or *needle bending* during the needle insertion process are much more significant and have the dominant impact on the overall accuracy of the needle puncture procedure. These issues are discussed later in section 9.1.

5.6 A Linear Stage for Translational Motions (CT table simulator)

The automatic CT-guided needle placement approach requires appropriate control of the horizontal CT table drive, in case that the needle is inserted tilted to the scan plane. The CT table, which has been used for experimental evaluation, provides a joystick that allows the radiologist to move the table during interventional procedures. This *joystick interface* was intended

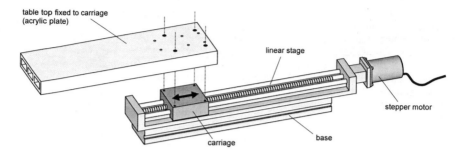

table top fixed to carriage
(acrylic plate)

linear stage

stepper motor

carriage

base

Figure 5.18:   Schematic of the 'CT table simulator' developed for the CT-guided needle placement experiments. It consists of a linear stage with a radiolucent acrylic plate attached to its carriage. A precise stepper motor is fixed directly to the screw shaft.

to be used by the robot control workstation to perform horizontal table motions by simulation
of pressing the joystick (see section 8.1).

There exist two joystick modes, 'continuous' and 'incremental' table motion. In *continuous mode* the CT table moves horizontally as long as the joystick is pressed. But there is no
way to get feedback of the table position. To get rid of this problem, the visual servoing
workstation would have to communicate with the internal CT CAN-bus system. However, to
realize this within an acceptable period of time was not possible for the author.

In *incremental mode* the CT table moves with predefined increments of 1 up to 10 millimeters, if the joystick is pressed. Therefore, the control workstation can determine the actual
table position by the number of moved increments. Although, this incremental mode is an

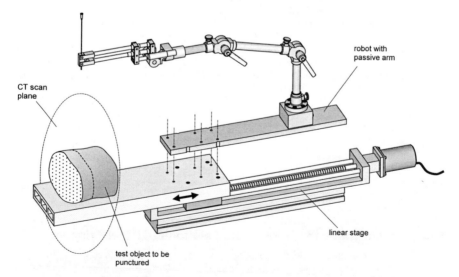

robot with
passive arm

CT scan
plane

test object to be
punctured

linear stage

Figure 5.19:   The 'CT table simulator' with attached passive arm and robot. The test object,
which is going to be punctured, is positioned on the acrylic plate. The linear stage
allows precise motion and positioning of the test object in the CT scan plane.

easy and convenient way to get access to horizontal table motion, it allows only for a poor control performance. In order to obtain an acceptable table positioning resolution, it is necessary to choose the smallest adjustable motion increment of 1 millimeter. However, a table translation of e.g. 5 cm would require the table to move with 50 increments, which leads to acceleration and deceleration of the table for 50 times (this would need about 50 seconds!).

In order to obtain a more appropriate horizontal motion performance, a special 'CT-table simulator' has been developed (see figure 5.18). This table simulator consists of a linear stage which allows to perform translational motions with high accuracy and excellent control performance. A radiolucent acrylic table top is fixed to the carriage and serves as support for positioning test objects which are going to be punctured (compare figure 5.19). Furthermore, the base plate of the passive arm is attached to the stage carriage as well. A precise 5-phase stepping motor (RDM 569/50, Berger Lahr Positec, Lahr, Germany) is fixed directly to the screw shaft of the linear stage. Together with the power electronics unit integrated in the visual servoing workstation (compare section 6.6.1) the stepper motor provides a resolution of 1000 steps per revolution.

The table simulator is placed on the unmoved CT table and allows to drive the test object within the CT scanner similarly to the CT table in common CT applications.

5.7   The 'Active Needle' – a Self-bending Needle

Precise manual needle placement requires much experience and high skills, since the physician has to deal with several handicaps: (i) movements of the patient or the target during the puncture, caused by patient breathing or due to the deformation of anatomical structures during the procedure; (ii) needle drift occurs if the needle is guided oblique through tough or stringy tissue; (iii) even the bevel grinded needle tip may lead to needle drift, especially along deep trajectories.

Already before needle insertion the experienced physician tries to take these factors into account and orients the needle slightly tilted with regard to the straight trajectory. Thus, the needle finally hits the target although the needle initially has been aimed aside it. However, desirable would be a kind of 'self-bending needle'. If needle drift occurs during insertion, the needle can automatically correct the direction of the further path by bending itself towards the

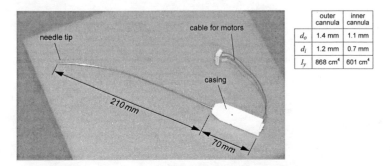

	outer cannula	inner cannula
d_o	1.4 mm	1.1 mm
d_i	1.2 mm	0.7 mm
I_y	868 cm⁴	601 cm⁴

Figure 5.20:   A photograph of the 'active needle'. It consists of (i) the two bent cannulas nested into each other and (ii) the drives for cannula rotation integrated in the white casing. The inner cannula runs through the casing and may admit a micro tube, micro catheter, or syringe at its proximal end behind the casing. The geometric parameters of the cannulas are listed in the small table on the right side.
(d_o, d_i: outer, inner diameter; I_y: geometrical moment of inertia)

target. Another benefit of such an 'active needle' would be the ability to puncture targets which are located behind bones or large vessels and that do not allow for a straight access trajectory.

A prototype of an 'active needle' has been developed for usage together with the needle guiding robot (see figure 5.20). Its mechanical principle is based on two bent cannulas which are nested into each other. If one needle is rotated relatively to the other, their elastic lines get superimposed and result in a new elastic line. In the ideal case this could lead to a neutralized deflection of both needles and to a straight line. There are several factors having influence on the resulting elastic line, e.g. the run of the initial elastic lines of both needles or their geometrical moment of inertia.

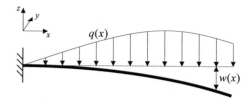

Figure 5.21: The elastic line $w(x)$ of a semibeam under a load distribution $q(x)$ along its axis. This model is used for the computation of the deflected cannulas of the 'active needle'.

From the general linearized differential equation, which describes the bending of a beam (compare figure 5.21)

$$w''(x) = -\frac{M_b(x)}{EI_y(x)}$$

$M_b(x)$ *bending moment along axis x*
E *elastic modulus*
$I_y(x)$ *geometrical moment of inertia* (5.21)

there results after double integration the elastic line of a semibeam

$$w(x) = -\frac{1}{EI_y}\iint M_b(x)dx^2 = -\frac{1}{EI_y}f(x).$$ (5.22)

The function $f(x)$ contains the load distribution $q(x)$ along the cannula. The combination of both cannulas results in a superposition of the elastic lines. The forces that both cannulas impose on each other (surface pressure) must be identical, except the sign:

$$q_1(x) = -q_2(x)$$ (5.23)

The elastic lines of cannula 1 and cannula 2 (see figure 5.22) can be formulated as

$$w_1(x) = -\frac{1}{E_1I_{y_1}}f_1(x)\ ,\quad w_2(x) = -\frac{1}{E_2I_{y_2}}f_2(x).$$ (5.24)

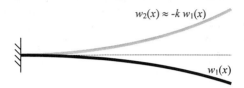

Figure 5.22:   Elastic lines $w_1(x)$ and $w_2(x)$ of the roundly bent shapes of the outer and inner cannulas.

It is assumed that both cannulas are consisting of the same material ($E_1 = E_2 = E$) and the relation between both elastic lines is $w_2(x) \approx -k\,w_1(x)$ (see figure 5.22). This is leading to $f_1(x) = -f_2(x)$. Thus, the elastic line of both cannulas can be written as

$$- EI_{y_1} \cdot w_1(x) = EI_{y_2} \cdot w_2(x) = -EI_{y_2} \cdot k\,w_1(x) \qquad (5.25)$$

In case that the resulting deflection of both cannulas gets neutralized, it must be

$$\frac{I_{y_1}}{I_{y_2}} = \frac{w_2(x)}{-w_1(x)} = k \qquad (5.26)$$

This result can be interpreted as follows: if both cannulas have the same elastic line ($k = 1$), than the geometrical moments of inertia I_{y_1} and I_{y_2} must be identical. This can be achieved by choosing cannulas with appropriate inner and outer diameter. The geometrical moment of inertia of a tube is defined as

$$I_y = \frac{\pi}{64}\left(d_o^{\,4} - d_i^{\,4}\right) \qquad \begin{array}{l} d_o \quad \text{outer diameter} \\ d_i \quad \text{inner diameter} \end{array} \qquad (5.27)$$

However, one is depending on available off-the-shelf cannulas for the construction of the 'active needle'. This could lead to $I_{y_1} \neq I_{y_2}$ and entails that one of the cannulas has to get a modified initial bending with the amount of $k = I_{y_1}/I_{y_2}$ (compare figure 5.22). This meets again equation (5.26). The geometrical parameters of the cannulas that have been used for the 'active needle' are listed in figure 5.20. Since the geometrical moments of inertia of the chosen cannulas are not identical, the inner cannula had to be bent for $I_{y_1}/I_{y_2} = 1.44$ times stronger

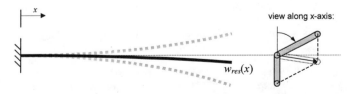

Figure 5.23:   The resulting elastic line $w_{res}(x)$ after putting the two roundly bent cannulas into each other.

than the outer cannula. The resulting elastic line $w_{res}(x)$, after putting the two roundly bent cannulas into each other, is approximately a vectorial addition of both elastic lines $w_1(x)$ and $w_2(x)$ as can be seen in figure 5.23.

The assembly of the active needle is shown in figure 5.24-a. The rotation of the inner and outer cannula is performed by two micro DC motors, integrated in the casing at the proximal end of the system (0.2W, gear ratio 256:1, FAULHABER GmbH, Schoenaich, Germany). Small gearwheels are fixed to the end of the cannulas. They allow to rotate the cannulas precisely and independently from each other. A deflection of the 'active needle' in one plane requires a simultaneous rotation of the inner and outer cannula in opposite directions.

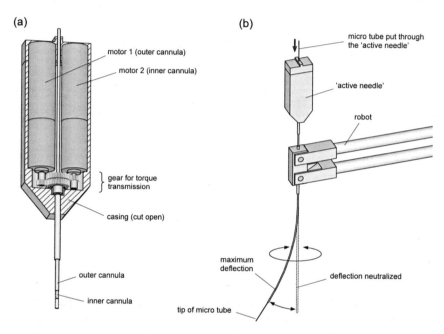

Figure 5.24: (a) The assembly of the 'active needle'. The rotation of the inner and outer cannula is performed by two micro DC motors. Small gearwheels are fixed to the end of the cannulas and allow to rotate the cannulas relatively to each other. (b) The 'active needle' as end-effector of the needle-guiding robot. a flexible micro tube is put through the inner cannula in order to puncture a target.

Although, the 'active needle' was designed as end-effector for the needle-guiding robot, it could principally be used manually as well. The deflection of the needle could be adjusted by pressing e.g. a button. However, orientation within the patient's anatomy is getting more challenging for the physician the more degrees-of-freedom he or she has to control. Therefore, the use of the 'active needle' together with a robot and a preoperative interventional planning is appropriate. The 'active needle' guided by the robot has been tested during a pig cadaver study using CT-imaging (see section 8.2.3). In these tests a micro tube ($\varnothing0.6$mm) has been inserted through the 'active needle' in order to puncture a target (compare 5.24-b). Alternatively, a syringe could by affixed to the inner cannula above the casing of the 'active needle'.

Chapter 6

Visual Servoing:
an Image Based Control for the Robot

6.1 General Considerations on Visual Control

Most robot applications require interaction of the robot end-effector with objects in the work environment. As already discussed in section 3.1, conventional robot control focuses on positioning the end-effector accurately in a fixed world coordinate frame. To achieve precise and reliable interaction with targets in the workspace, these objects have to be consistently and accurately located in the same world coordinate frame. Uncertainties in the object position or the end-effector position will lead to a position mismatch and often to failure of the operation.

Beside the use of precise joint angle encoders, '*vision*' provided by a camera got increasingly attention in the manufacturing field in the last decade. Visual control of manipulators promises substantial advantages when working with targets whose *position is unknown*, or with manipulators which may be *flexible* or *inaccurate*. It can support accurate free-space motion in environments that are less structured than those found with today's industrial robots. The reported use of visual information to guide robots or mechanisms, is quite extensive and encompasses manufacturing applications, remote teleoperation, missile tracking cameras, fruit picking, nuclear waste cleanup, intelligent highway systems as well as robotic ping-

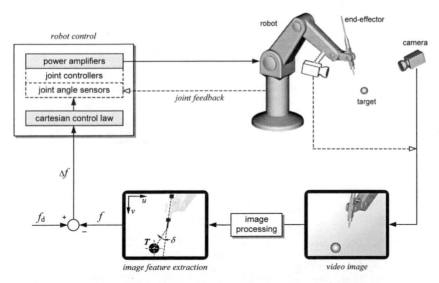

Figure 6.1: Controller block diagram of a visual-feedback robot control ('dynamic look-and-move' control). Depending on whether the control loop incorporates joint feedback information or not, the type of control is called a 'dynamic look-and-move', otherwise 'image based servoing' in case that no joint feedback or joint controllers are employed. The camera could be either stationary fixed or end-effector mounted.

pong, juggling, and balancing. An excellent overview of the main issues in visual servo control of robot manipulators, and a good review of the active research in this area, is given by Corke [33].

There are several types of visual control techniques, like stationary or end-effector-mounted cameras (compare figure 6.1), monocular or binocular vision, planar or complete 3D motion control.

Traditionally visual sensing and robot manipulation are combined in an open-loop-fashion, 'looking-then-moving', i.e. the camera takes only one image which is analysed and used to move the end-effector to the desired position. In this case the 'image based control-loop' is performed only once, so the accuracy of the operation depends directly on the accuracy of the visual sensor as well as the manipulator itself. An alternative to increase the accuracy is to 'close' the loop by a *visual-feedback control loop*, which is shown in principle in figure 6.1. If the robot system incorporates vision on base of task level programming using joint feedback information, it is called a *dynamic look-and-move* control. If servoing is done only on the basis of image features without joint feedback and joint controllers, *image based servoing* is obtained as proposed by Weiss [171].

The following sections will give a brief overview of the basic issues of visual servoing, relevant for this thesis.

Position-based Servo Control (dynamic look-and-move)

In the case of *position based* servo control, the features measured in the image plane are used to explicitly calculate the 3D-pose of the end-effector and the target at any time. To achieve this, the geometry (3D model) of the target object and its features must be known [166] (see figure 6.2-a). The 3D position of the end-effector and target can be estimated using either active or passive techniques. Structured light projection systems and optical navigation systems

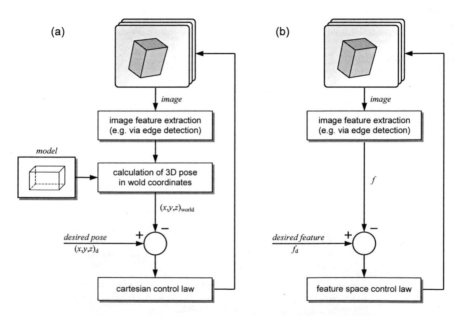

Figure 6.2: Controller block diagram of two different types of visual control techniques, (a) position-based servo control, (b) image-based servo control.

are two examples of active range sensors. Passive techniques include *photogrammetry* [177], *stereo vision* [178] and *depth from motion* [111]. The control design is then based on the difference between the estimated and the desired pose of the end-effector in world-coordinates. This deviation is used to calculate the control error function [173].

Image-based Servo Control

By contrast to position-based servoing, in *image-based* visual control the elements of the task are specified directly in the *image feature space*, not in world space (see figure 6.2-b). Mostly the real world task is described by one or more image space tasks [173]. For instance, the robot motion is controlled to achieve *desired conditions in the image*, e.g. the alignment of a tool with a target, as shown in figure 6.1. This can be done without knowing or computing the real world coordinates at any time during the execution of the task. The control feedback loop directly uses the location of features on the image plane. Therefore, it is possible to have successfully executed the task without finally knowing the 3D coordinates of the target in the world coordinate frame.

In this thesis an image based servo control has been implemented for the presented needle-guiding robot in a dynamic look-and-move fashion (compare figure 6.1). The scene in the digital images is analysed via image processing algorithms to extract the image features f, like the target position f_T, the needle pose f_N, and the needle deviation angle f_δ in the image. In consideration of all actual image features $f(f_T, f_N, f_\delta)$ and the desired image features f_d, the 'feature space control law' employs $\Delta f = f_d - f$ to compute the input parameters for the joint controllers of the robot (figure 6.2-b).

In preliminary experiments a digital video camera has been employed to develop and evaluate the new visual control algorithms. For the ensuing cadaver studies different medical imaging modalities, as X-ray fluoroscopes or a CT scanner, have been used for image acquisition in the visual servoing control for the robot.

6.2 Visual Servoing in Medicine

A central concern of visual servoing research is the dynamic issue of visual control, since the economic justification is frequently based upon cycle time. High sample rate for the vision sensor and robot control, and low latency and high-bandwidth communications are critical to a short settling time [34]. However, in medicine, the situation is radically different. In X-ray or CT medical imaging for example, high imaging sample rates - which would be desirable in industrial robot applications - would result in high radiation exposure for the patient. Furthermore, guidelines for industrial robots suggest that a robot should not be powered when people are in its vicinity. This would be inappropriate in the OR, where robots are working very closely together with surgeons and the patient [43]. Therefore, medical robots are generally moving very slowly, and sophisticated dynamic visual control issues are of lower importance in the medical field.

Since the use of medical robots has just come up during the last decade, there are currently not many visual servoing systems developed for medical purpose. Most of these systems that are using a visual control have been proposed in the field of minimally invasive surgery to control the motion of a camera or an endoscopic instrument [158][28][95].

A system, recently developed, is a laparoscopic guiding system for an AESOP 1000 robot (Computer Motion, Inc., see figure 3.7-a), developed by the German Aerospace Research Establishment (DLR) [170]. Using color image segmentation the system automatically locates the tip of a laparoscope provided with a green mark. This image feature (green mark) controls the laparoscopic camera so that the instrument's tip is always at a certain position in the image, e.g. the image center.

Salcudean *et al.* [131] proposed a special counterbalanced robot for positioning an ultrasound probe. They used visual servoing for robot image feature tracking in the ultrasonic image plane.

To the author's knowledge, the actual thesis presents the first experimental setup and results on visual servoing using either X-ray fluoroscopy or CT-imaging. The original ideas and theoretical framework of the X-ray fluoroscopy approach was first presented by two researchers of the Siemens Corporate Research, Princeton, USA [110]. This method is presented in the following section 6.3. After the successful implementation and evaluation of this approach the author transferred these ideas into CT-based servoing and focussed on automatic CT-guided needle placement, which is described in section 6.4.

6.3 Visual Servoing Approach for X-ray Fluoroscopy

This section describes the principle of the new image-based control for semi-automatic and uncalibrated needle positioning using *X-ray fluoroscopy* provided by a uniplanar X-ray C-arm (see figure 6.3). First, the user has to identify the lesion in the X-ray image and must choose an appropriate insertion point for the needle. Then, the passive arm with the robot is manually moved in order to place the tip of the guiding cannula at the chosen insertion point on the patients skin. During the automatic alignment process the robot rotates only the needle around the insertion point. The image feature extraction requires, that the target, the skin entry point

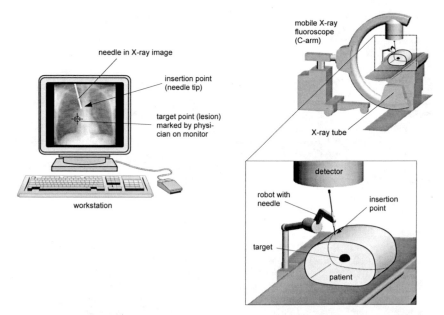

Figure 6.3: Needle placement setup with robot and uni-planar mobile X-ray fluoroscope (C-arm). First, the physician places the needle tip at the chosen insertion point on the patient by moving the robot manually. Now, the C-arm is positioned in a way that the target point and at least the needle's tip are visible in the X-ray images. Before starting the automatic alignment procedure, the user has to define the target point on the displayed radiograph on the workstation monitor.

and the distal end of the needle are visible in the image (compare figure 6.3). Then, the physician defines the target point by clicking on the X-ray image on the monitor. The goal of the following alignment process is to rotate the needle around the fixed skin entry point to achieve alignment with the target structure. This is performed by the robot from two arbitrary and uncalibrated X-ray views using a novel visual servoing technique.

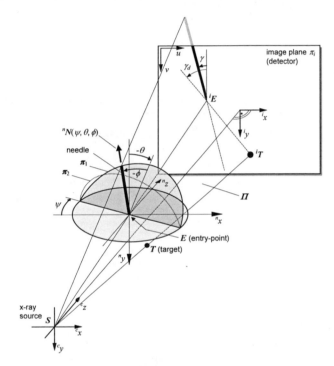

Figure 6.4: The projection geometry can be approximated by a pinhole camera model (X-ray source as the optical center S, detector as image plane π_i). The needle pose ${}^nN(\psi, \theta, \phi)$ in frame n is given by the Euler-angles ψ, θ, ϕ.

6.3.1 Method for Uncalibrated Needle Placement

The projection geometry of the X-ray fluoroscope is given approximately by a pinhole camera model, with the X-ray source as optical center S, and the detector as image plane π_i (see figure 6.4). E is the skin entry point of the needle on the patient's body and the target point (lesion) is noted by T. ${}^iE$ and ${}^iT$ are their radiographic images. The three-dimensional position of T is unknown. The only information available about T is the 2D position of its projection ${}^iT$ in the radiographic image, which has been specified by the surgeon via mouse-click on the monitor. Additionally, Π is defined as the viewing plane containing the optical center S, the target on the image ${}^iT$, and the fix point E. The maximum information one may get from this one view is the three dimensional position of the plane Π related to the needle coordinate frame n. The needle pose ${}^nN(\psi, \theta, \phi)$ in frame n is given by the Euler-angles ψ, θ, ϕ (compare section 5.3.1). In order to get the complete 3D information about the desired alignment pose of the needle, the system requires two arbitrary, non-identical views of the scene. A detailed description of the X-ray guided alignment process is given in the following four paragraphs:

Step I

The robot rotates the needle in an arbitrary plane π_1, defined by ψ_1 and θ_1, around the fixed point E (compare figure 6.4). A good choice for π_1 would be a plane approximately perpendicular to the viewing direction. The actual needle orientation in the image is given by the angle γ between the projected needle axis and the vertical. The desired needle orientation in the image is γ_d, which represents the alignment of the projected needle with the image target point ${}^iT_1$, i.e. the virtual axis of the projected needle goes through the image target point ${}^iT_1$. The deviation angle between the projected needle in the image and the line $\overline{{}^iE_1\ {}^iT_1}$ is given by δ. For the visual servoing alignment process one can define the image feature error function as $\delta = \gamma_d - \gamma$ (deviation angle), which vanishes in case of alignment in the image. After this first alignment step, the corresponding needle pose in frame n is ${}^nN_1 = {}^nN(\psi_1, \theta_1, \phi_1)$, which represents the three-dimensional intersection of plane π_1 with the viewing plane Π_1 (see figure 6.5-a).

Step II

The system chooses a second plane π_2 passing through E but different from π_1 (ψ could be kept constant). Now the robot rotates the needle around E in plane π_2 till the projected needle and ${}^iT_1$ are again visually aligned in the image (see figure 6.5-a). This final position of the needle in frame n is given by a second three-dimensional direction vector ${}^nN_2 = {}^nN(\psi_1, \theta_2, \phi_2)$ which gives the system the intersection of plane π_2 with the viewing plane Π_1. The two distinct vectors ${}^nN_1$ and ${}^nN_2$ uniquely define the desired viewing plane Π_1 in the needle coordinate frame n. This is the maximum information that can be determined from a single viewpoint with no prior calibration data. Note, that plane Π_1 had to be defined with the two needle

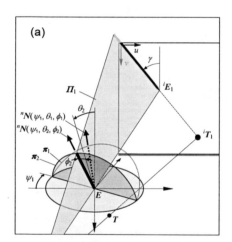
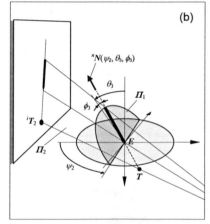

Figure 6.5: The X-ray guided alignment process requires two arbitrary and uncalibrated X-ray views for complete 3D alignment. In both views the needle is moved around its entry point till target alignment in the radiograph is obtained. (a) Searched first viewing plane Π_1 after 2D needle alignment in the first X-ray view. (b) After a second 2D needle alignment while moving the needle only in plane Π_1 in a second arbitrary viewpoint, the final 3D alignment of needle is obtained.

orientations $^nN_1$ and $^nN_2$, since the optical center S is unknown. The next step is to use a second viewpoint.

Step III

To observe the needle alignment from another viewpoint, the X-ray C-arm is rotated to an arbitrary different orientation (see figure 6.5-b). Obviously, a rotation of about 90° will result in most accurate placement of the needle, but this might be limited e.g. by equipment in the operating room. A new X-ray image is taken, and the physician is asked again to identify the position of the target in the new image on the monitor. This 2D position is called $^iT_2$. The new viewing plane is Π_2, and is defined by the new position S_2 of the X-ray source, the target T and the skin entry point E.

Now the robot moves the needle only in the obtained first viewing plane $\Pi_1(\psi_2, \theta_2)$, which is described in the needle coordinate frame by the new Euler-angles ψ_2 and θ_2. Again the image feature error function $\delta = \gamma_d - \gamma$ can be computed, and the needle is moved till δ vanishes. The needle orientation $^nN_3 = {}^nN(\psi_2, \theta_3, \phi_3)$ that is obtained now, is the desired 3D needle pose that aligns the needle with the target point in 3D. The last step of needle placement procedure is then to measure the required insertion depth of target T relative to entry point E.

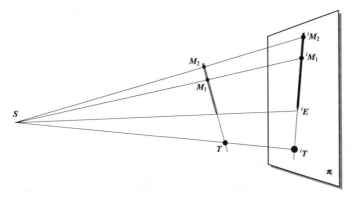

Figure 6.6: After the 3D alignment of the needle, the required insertion depth can be estimated using cross-ratios between characteristic points on the needle (needle tip and at least two additional markers M_1 and M_2), the target point and their radiographic projections in the image plane π_i.

Step IV

Since cross-ratios are invariant under perspective projection, they can be used to calculate distances in 3D out of their projected distances on the image plane. However, to obtain Euclidean measurements of 3D distances, reference 3D distances are needed. Therefore, it is assumed to have two metallic marker beads, M_1 and M_2, perfectly located on the needle axis and with known distances (see figure 6.6). The cross ratio $[^iM_1, {}^iM_2, {}^iE, {}^iT]$ of the image points (image coordinates) will be computed from the digital image after the needle has been aligned with the target. This cross-ratio can be derived as a ratio between relations of distances between these points as follows:

$$\left[^iM_1, {}^iM_2, {}^iE, {}^iT\right] = \frac{\left\|^iE \, ^iT\right\|}{\left\|^iM_2 \, ^iT\right\|} : \frac{\left\|^iE \, ^iM_1\right\|}{\left\|^iM_2 \, ^iM_1\right\|} = \lambda = const. \tag{6.1}$$

According to projective geometry, the computed cross-ratio λ of the image points is identical to the cross-ratio $[M_1, M_2, E, T]$ of the corresponding three-dimensional points in the world coordinate frame. Since the Euclidean distances $\|M_2 M_1\|$, $\|EM_1\|$ and $\|M_2 T\|$ on the needle are known, the insertion depth $\|ET\|$ can be easily computed by the following equation:

$$\lambda = \frac{\|E\,T\| \cdot \|M_2\,M_1\|}{\|M_2\,T\| \cdot \|E\,M_1\|}.$$

(6.2)

Substitution with $\|M_2\,T\| = \|ET\| + \|EM_2\|$ finally leads to the desired insertion depth of the needle:

$$\|E\,T\| = \frac{\lambda \cdot \|E\,M_1\| \cdot \|E\,M_2\|}{\|M_2\,M_1\| - \lambda \cdot \|E\,M_1\|} \qquad \text{with} \qquad \lambda = \frac{\|{}^iE\,{}^iT\| \cdot \|{}^iM_2\,{}^iM_1\|}{\|{}^iM_2\,{}^iT\| \cdot \|{}^iE\,{}^iM_1\|}.$$

(6.3)

The use of these marker beads can be realized by employing (a) a special needle phantom during the alignment procedure, which is provided with beads, or (b) with the guiding cannula provided with beads. Once the automatic alignment procedure is finished, the needle has to be placed in the guiding cannula and the puncture of the target can be performed. In practice more markers can be used and the depth is estimated in a least-square sense in order to increase precision (compare section 7.5.1).

6.3.2 Point Reconstruction and Metric Measurement

Performing needle placement and estimating the insertion depth, completely reconstructs the three-dimensional position of the target point regarding to the needle coordinate frame N. Thus, by successively 'targeting' two or more 3D points inside the patient using the presented method, allows to determine metric distances and orientations between anatomical structures. This approach could be used for instance to estimate the distance between a tumor to be removed and the nearest major blood vessel or nerve prior to surgery, without an expensive and time consuming tomographic reconstruction of the whole volume.

6.3.3 Visual Servo Control Law

In this medical application the image feature to control is the angle γ between the current needle orientation in the image and the y_i-axis of the image plane π_i. Together with the desired image feature γ_d (angle between line $\overline{E_i T_i}$ and the y_i-axis) the image feature error function $\delta = \gamma_d - \gamma$ can be derived. This error function (or deviation angle) vanishes in case of perfect needle alignment with the target. Since a projected needle in the image consists of a finite number of discrete pixels, the needle orientations that were determined in the image are discrete as well. This effect can get significant if the needle consists only of a small number of image pixels. Therefore, a small interval around the desired image feature $\gamma_d \pm \varepsilon$ must be admitted. For the needle alignment experiments a value of $\varepsilon = 0.2°$ turned out to be appropriate and led to the *stopping criterion* of the visual servoing feedback loop: $|\delta| = |\gamma_d - \gamma| < 0.2°$.

Since the relation between changes in the image feature γ to changes in the angle ϕ of the needle in plane π is unknown, a *proportional control law* has been implemented as an approximation. That means, if the image feature error function $\delta = \gamma_d - \gamma$ is determined, then the robot control moves the needle relatively in plane π by the magnitude $\phi = \delta$. An alternative would be to use an *adaptive control* approach where the system 'learns' the relationship be-

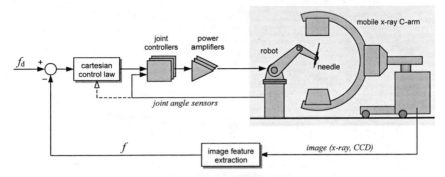

Figure 6.7: Controller block diagram of the visual servoing feedback loop for automatic needle alignment using X-ray fluoroscopy guidance.

tween image feature motion and joint-space motion after a number of trajectories [110]. Thus, the control could estimate the function $\Delta\phi = f(\Delta\delta)$ during the alignment process. But it can be shown that this would require already 3 alignment steps. Therefore, even with an *adaptive control*, at least 4 loops for alignment in the image are needed. As the experimental results showed, the implemented proportional control law typically needs only 4 to 5 alignment steps anyway (see chapter 7).

The controller block diagram is shown in figure 6.7. The visually controlled needle alignments in plane π_1 and π_2 can be understood as *image-based servoing* since the image feature (angle γ) to control the robot is directly used without computing geometrical relationships. Then, when deriving Π_1 or the reconstructed target pose, this approach can be seen as *position-based servoing*, employing the *depth from motion* approach. Note, that using depth from motion, one has to assume that the target pose is constant between the views. Therefore, patient or organ movement, and thus target motion during the servoing procedure can cause errors in needle alignment.

6.3.4 Workflow and User Interface

While paragraph 6.3.1 gave a more theoretical overview of the X-ray fluoroscopic guided needle placement technique, the following section describes the workflow of this novel approach by means of a flow chart and presents the user interface.

The mobile X-ray C-arm used in the needle placement experiments offered several degrees of freedom to move and rotate the X-ray tube and intensifier around the patient. This allowed convenient imaging from different viewpoints. Figure 6.8 gives an overview of the *workflow* of the novel X-ray fluoroscopy approach. A more detailed flow chart of the needle placement algorithm is shown in the appendix of this thesis (see page 177).

During initial diagnostic imaging, the anatomical region of interest is visualized. The target area has to be identified and the ensuing intervention is planned. This implies the definition of the most appropriate access trajectory and the resulting insertion point, which is marked on the patients skin. Then, the manipulator has to be installed on the patient. The tip of the guiding cannula, which is part of the manipulators end-effector, must be located on the marked insertion point. During the alignment procedure, the manipulator rotates the guiding cannula around its tip to achieve the desired orientation. The puncture needle might already be put in the guiding cannula before the procedure starts, or after alignment of the guiding cannula is achieved.

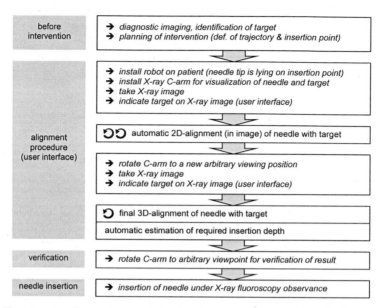

Figure 6.8:   Workflow of new approach to automatic X-ray fluoroscopy guided needle alignment.
The circular arrow symbolizes activation of the visual servoing control loop (↻).

When the manipulator has been installed at the insertion point on the patient, the user interface and control software on the host computer has to be started. The layout of the user interface for X-ray fluoroscopy guided needle placement is shown in figure 6.9. The user is guided
through the procedure by instructions automatically displayed by the interface (2). The current X-ray image is always shown on the screen (1). Successful needle alignment requires the
visibility of the target as well as the distal end of the needle in the image at the same time. If
not, the C-arm has to be repositioned to achieve this necessity.

Image acquisition is triggered by the host computer via I/O interface card. The trigger signal is send to a special switch box, which is connected to the cable of the C-arm foot pedal
(see section 7.2). The user can e.g. manually press a button on the user interface (6) to acquire
a single fluoroscopic image.

After starting the user interface the manipulator gets initialized and the user is asked to install the manipulator on the patient and to position the X-ray C-arm. Then, one single image
from the fluoroscope is acquired and displayed on the computer screen.

Now, the physician can position the target-marker (small crosshair) by clicking with the
mouse at the desired location on the image. By confirming this position with the 'fix target'
button, the target point is defined and displayed in the target list (5). The user can define up to
five target points in the image (multi target alignment). Later, after the alignment process has
finished for all these target points, this approach allows to determine metric distances and
orientations between anatomical structures (see paragraph 6.3.2). A special functionality is
integrated for targeting metal beads. In case of high contrast targets, that show up sharply
bounded in the image, the center of gravity can be precisely determined if 'catch point' is activated (11). The user has just to click somewhere inside the bead's contour in the image and
its center of gravity is automatically determined. In medical practice this functionality would

be of lower interest, but for precise verification of the principle accuracy of this alignment technique, the accurate target definition in both X-ray views is essential.

After all target points are indicated by the physician, the host computer automatically starts the first visual servoing control loop. The computer selects a first plane π_1 to move the needle within (see paragraph 6.3.1, Step I). The robot changes the needle orientation automatically till the image shows alignment of needle and target. This stepwise alignment process can be observed by the user on the screen. If several targets have been marked, the computer automatically repeats this alignment procedure for each target sequentially. When finished, the computer automatically selects a different plane π_2 to move the needle in and the alignment procedure is repeated for each target point. The alignment control loop for both planes is performed automatically by the computer (symbolized with ⊃⊃ in figure 6.8).

For the next alignment step, the user is asked to rotate the C-arm to a new arbitrary view-

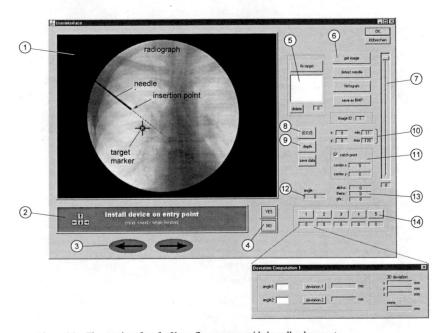

Figure 6.9: The user interface for X-ray fluoroscopy guided needle placement.
(1) display of video image (X-ray or CCD)
(2) instructions for user during the needle placement procedure
(3) back/next buttons to navigate through the menu
(4) answering to questions asked by the user interface
(5) confirmation and display of target points (image coordinates)
(6) buttons for manual release of e.g. image acquisition or needle detection
(7) slider for remote controlled needle insertion (only for LARS and SIEMON)
(8) button to move needle back to initial orientation
(9) button for automatic depth computation (required insertion depth)
(10) grayscale values min/max used as thresholds for needle segmentation in image
(11) image coordinates of current target (detected center of gravity of target bead)
(12) current deviation angle between needle axis and target point
(13) Euler-angles of current needle orientation (end-effector coordinate system)
(14) automatic computation of deviation between needle axis and target (millimeter)

ing position. Again, the visibility of the target and the distal end of the needle in the image is required. Then, one single image is acquired and displayed on the user interface in order to define the target points in this second view. If more than one target has been chosen in the first view, it is required to follow the same sequence of target marking in the second view. Occasionally, in fluoroscopic imaging a target structure may be difficult to visualize, for example if other structures, like bones, are overlaid with the target. In this case, the physician has to modify the viewing position of the X-ray C-arm in order to obtain a more appropriate viewpoint.

After indication of all targets in the second view, the host computer starts again the visual servoing control procedure for final needle alignment in the image. However this time, the *desired 3D alignment of the needle with the target* has been achieved (compare paragraph 6.3.1, Step III). The guiding cannula is provided with marker beads, which are used for esti-

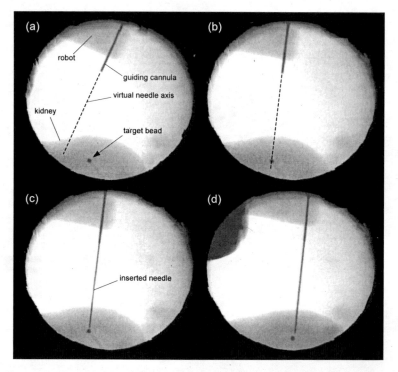

Figure 6.10:   Needle alignment experiment with an X-ray fluoroscope performed at the Johns Hopkins University with the RCM robot. A ∅2mm metal target bead has been implanted into a single pig kidney. (a) Starting position of the robot. The virtual needle axis is drawn onto the image. (b) Final orientation of the guiding cannula after the complete alignment procedure. (c) Needle is inserted through the guiding cannula and placed close to the target bead. Here the insertion depth was visually controlled by the physician looking on the screen. (d) Verification of the puncture from a different viewing direction. The frayed image borders are caused by the distortion correction algorithm.

mation of the required insertion depth of the needle[20]. This automatic depth estimation is the last step of the visual servoing control, which leads to the complete reconstruction of the three-dimensional position of the target point with respect to the needle coordinate frame n.

Now, the automatic alignment procedure is finished and the user is asked to rotate the C-arm to acquire images from several viewpoints in order to verify the suggested needle orientation and insertion depth in these views.

After the physician made sure that the needle is correctly aligned, the robot is locked in its position. Then, the needle can be inserted manually or remotely controlled with the slider on the user interface (7). The actual insertion depth of the needle is always displayed.

During the alignment process all functions - like image display, needle detection, computation of the deviation angle - are performed automatically during the control loop. However, the user interface provides several buttons for manual execution of these functions (6)(9). Furthermore, parameters of the current status of e.g. the robot (Euler-angles) are displayed (12)(13). The needle is detected by a simple threshold algorithm, where the needle pixels are identified as grayscale values between a lower and upper limit (10).

Figure 6.10 shows four X-ray images taken during the automatic needle alignment procedure performed at the Johns Hopkins Medical School with the RCM robot. As target served a $\varnothing$2mm metal bead which has been implanted into a single pig kidney. In this experiment the tip of the guiding cannula (insertion point) is located about 6 centimeters above the kidney, without further tissue in between the kidney and the 'virtual' insertion point. Below the target bead a thin wire is slightly visible, which has been soldered to the bead. It allows automatic electrical contact sensing to detect contact between the needle and the metallic target bead during needle insertion. The radiolucent robot end-effector with its guiding cannula shows up in the upper image area. After the automatic alignment procedure the puncture needle is inserted through the guiding cannula into the kidney and is placed close to the target bead (figure 6.10-c). During needle insertion the insertion depth was visually controlled by observing the fluoroscope screen. After needle insertion, the C-arm is rotated to obtain a different viewing direction to verify the accurate placement of the needle (figure 6.10-d).

6.3.5 Visual Servoing with the 'Axial Aiming Technique'

Another approach of automatic X-ray guided needle placement using visual servoing has been developed and tested. As discussed in section 2.2, the *axial aiming technique* is a well established manual needle guidance method for X-ray fluoroscopy. In this approach the C-arm is moved as long as the needle insertion point and the target are superimposed in the radiographic image. Then, the needle is rotated manually around the insertion point while imaging, till the needle degenerates to a point in the radiograph. This automatically leads to three-dimensional alignment of the needle with the target.

The same procedure has been adopted for automatic visually controlled needle alignment performed by a robot. The robot rotates the needle stepwise around its two main axes with $\Delta\alpha_i$ and $\Delta\beta_i$ (motor angles; compare section 5.3.3). The resulting motion of the needle in the images is detected and used for computation of the next motion increments $\Delta\alpha_{i+1}$ and $\Delta\beta_{i+1}$. Goal is to rotate the needle in a way, that the needle length $|n_i|$ in the image gets minimized. Therefore, this process is repeated till the needle degenerates to a point in the radiograph.

The principle of the developed algorithm is demonstrated in figure 6.11. If the robot rotates the needle with angles $\Delta\alpha$ and $\Delta\beta$, the needle hub moves on circular lines as shown in figure 6.11-a. First, the robot automatically rotates the needle with a predefined increment of

---

[20] In the experiments at the Johns Hopkins University a special needle phantom has been used, to which the marker beads have been fixed. In medical practice these markers would be integrated in the guiding cannula.

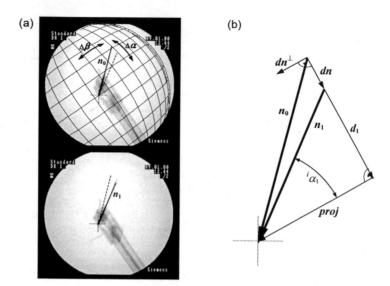

Figure 6.11: The robot rotates the needle around two perpendicular axes with angles $\Delta\alpha$ and $\Delta\beta$.
(a) The two radiographs show the needle before (n_0) and after (n_1) rotation with $\Delta\alpha_0$.
The circular lines drawn on the image above symbolize the sphere that would be described by the needle hub when rotating with $\Delta\alpha$ and $\Delta\beta$. (b) The right image shows the
analysis of both needle poses in the image and the determination of angle ${}^i\alpha_1$.

$\Delta\alpha_0 = 5°$ around its first motor axis. In the image the needle moves from its initial pose n_0 to
n_1. The difference vector dn in the image can be easily determined:

$$dn = n_1 - n_0 \ . \tag{6.4}$$

Now, the image needle vector n_1 is projected on the orthogonal vector $dn^\perp$

$$proj \ = \ \frac{n_1 \cdot dn^\perp}{\left\| dn^\perp \right\|^2} dn^\perp , \tag{6.5}$$

and the image deviation angle ${}^i\alpha_1$ between vector $proj$ and n_1 can be computed as

$${}^i\alpha_1 \ = \ \arctan\left(\frac{|proj|}{|n_1|} \right) . \tag{6.6}$$

The algorithm assumes, that a needle pose $proj$ in the image represents the final motor angle α. Certainly, angles measured in the radiograph do not directly allow to determine any 3D
deviation angle in the robot coordinate frame, i.e. rotating the needle with ${}^i\alpha_1$ will not lead to
the needle pose $proj$ in the image. However, the goal of the alignment algorithm is to advance
the needle stepwise towards the desired needle pose. For this purpose, a proportional control

law for the rotational increments of the motor axes has been defined. The next rotational increment for the first motor axis is computed as

$$\Delta\alpha_1 = k \; {}^i\alpha_1 = k \arctan\left(\frac{|proj|}{|n_1|}\right). \tag{6.7}$$

The practical verification of the algorithm showed that a parameter $k=0.7$ leads to good results and a stable alignment procedure. The alignment typically needed 7 to 9 iterations (compare figure 6.12). Needle rotations around the two perpendicular axes with angles $\Delta\alpha_i$ and $\Delta\beta_i$ are performed in an alternating manner. For this purpose, the presented approach for computation of the rotational increments $\Delta\alpha_i$ is applied for $\Delta\beta_i$ in the same way. Figure 6.12 shows a sequence of X-ray images taken during the automatic alignment procedure (without a target bead).

In case that a needle rotation does not lead to a shortening of the needle length in the image, $|n_{i+1}|>|n_i|$, the sign of the angle is switched in order to move in the opposite direction for the next needle motion.

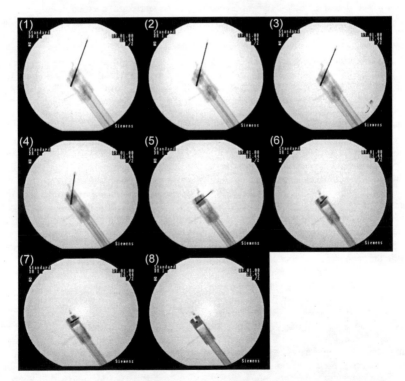

Figure 6.12: Sequence of radiographs taken during the automatic needle alignment procedure along an X-ray beam. Figure (1) shows the initial pose of the needle. The robot rotates the needle stepwise with certain increments around its two main axes (motor angles $\Delta\alpha$ and $\Delta\beta$). This process is repeated till the projection of the needle in the image degenerates to a point (8).

6.4 Visual Servoing Approach for CT-imaging

This section describes the basic principle and implementation of the new visual servoing approach for automatic and uncalibrated needle placement using CT-imaging. As presented in section 2.3.1, CT-guided needle placement can be performed in basically two ways, needle placement *within the CT scan plane* or *oblique to the scan plane*. Therefore, the following two scenarios have been defined for automatic image-guided needle placement with CT-imaging:

a) *The target structure and the needle insertion point are located <u>in the same CT scan plane</u>. Therefore, the access trajectory for the needle and thus the complete insertion procedure is visible in one image plane.*

b) *The target and the insertion point are located <u>in two different CT scan planes</u>. In this case, the access trajectory and thus the insertion procedure is not visible in the same image plane.*

For both scenarios the target structure has first to be placed within the image plane by moving the CT table. Now, the operator has to identify the target structure in the CT image displayed on the monitor (compare figure 6.13). After manual placement of the robot at the desired insertion point, the automatic alignment procedure can be started. A detailed description of the needle placement workflow for both scenarios is presented in the following two paragraphs.

Scenario (a): needle placement within the scan plane

First the robot has to be placed manually at the desired insertion point on the patients skin. The tip of the guiding cannula (rotation point) is now identical with the insertion point. In the next step the CT table is moved automatically while imaging, until the tip of the guiding cannula is visible in the image (the needle might be already placed in the guiding cannula). Using image processing, the tip of the cannula is automatically detected in the image (see figure 6.14-a). Then, the robot is tilting the guiding cannula as long as it appears with maximum length in the CT image. But the robotic system still does not know the orientation of the CT scan plane in the robot's coordinate system. Therefore, a second needle position in the CT scan plane is needed in order to define its position and orientation in the robot's coordinate frame. So the robot rotates the needle around its first rotation axis (compare figure 5.3) with a certain amount and tilts the needle again into the scan plane until it gets visible with maximum length (figure 6.14-b). Now, the CT scan plane is defined in the robot's coordinate frame by the two needle orientations. This registration between the robot and the CT scan plane is part of the automatic alignment procedure. The next step is to rotate the needle within this plane until alignment with the target is achieved. This is done by rotating the needle as long in the scan plane as the measured deviation angle in the image vanishes (see figure 6.14-c). The whole procedure is done automatically while the needle insertion itself can be performed manually or remotely controlled by the physician.

Scenario (b): needle placement oblique to scan plane

In this second scenario the target structure and the insertion point are not located in the same CT image plane. First, the target structure has to be located in the CT scan plane by movement of the table top. Then, the physician chooses the target point in this image plane (image coordinates x_1, y_1). After defining the best access trajectory, the robot's guiding cannula is placed at the resulting insertion point on the patient's skin (is not located within the actual scan plane!). The physician moves the CT table until the needle tip (insertion point) becomes visible in the image (see figure 6.15-a). Then the system precisely adjusts the CT table automatically while imaging, in order to guarantee precise positioning of the guiding cannula tip in the CT image plane. At this point, the image coordinates of the target (x_1, y_1) and the inser-

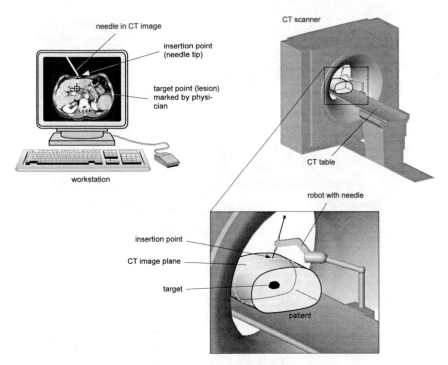

Figure 6.13: Needle placement setup with robot and CT scanner. Insertion of needle in image plane: first, the physician places the target within the CT scan plane by moving the CT table top. Then, the needle tip is located in the CT scan plane at the appropriate insertion point on the patient. Now, the needle tip and the target point should be visible in the CT image. Before starting the automatic alignment procedure, the user has to define the target point on the CT image.

tion-point (x_2, y_2) are known for the system. Since the CT table translation ΔT between the *target* image plane and the *insertion-point* image plane is known, the complete three-dimensional settings are determined in the CT image space. In the next step, the robot has to be registered to the CT scan plane, as described in scenario 1 (see figure 6.15-b). After the scan plane is known in the robot's coordinate frame, the robot can automatically adjust the needle orientation which aligns the needle with the target. In this second scenario, needle insertion has to be performed remotely by the physician. While in scenario 1 the needle insertion and deviations can be directly observed in the image, this is not possible in the second scenario. However, during needle insertion the system is automatically tracking the needle's tip by moving the CT table during imaging. Thus, the needle tip is always visible as a bright dot in the image. Also projected onto the image is the desired insertion trajectory (figure 6.15-c). This allows the physician, even in case of needle insertion tilted to the scan plane, a kind of visual monitoring of the needle path during remotely controlled insertion.

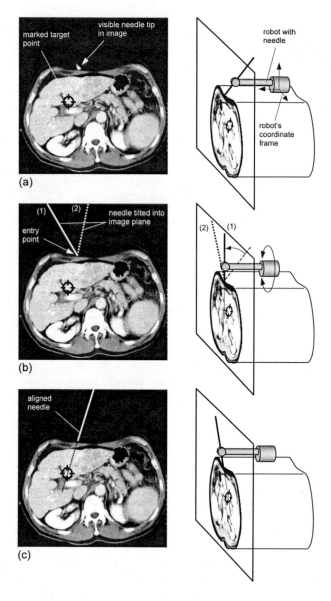

Figure 6.14: The automatic alignment process in case that insertion point and target are located within the same CT-image plane (scenario 1).

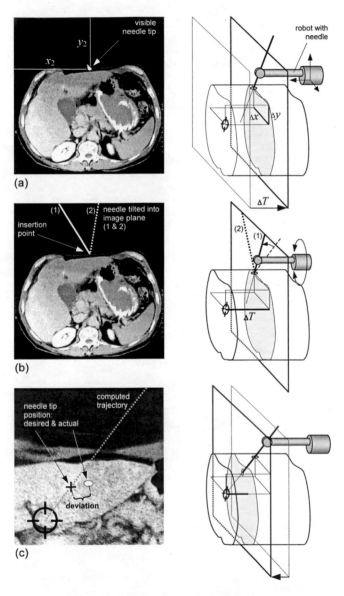

Figure 6.15: The automatic alignment process in case that insertion point and target are located in
two different CT-image plane (scenario 2).

6.4.1 Visual Servo Control Law

The controller block diagram of the CT-guided visual servoing feedback loop is shown in figure 6.16. Using an *image based control*, the elements of the task are specified in the *image feature space*. In this CT application there are two image features to control:

The first image feature is the *visible needle length L* in the image, which is needed to control the automatic tilting of the needle into the CT scan plane (see figure 6.17-a). In case the needle is completely located within the scan plane, L increases to a maximum. The diagram in figure 6.17-b shows that close to the desired needle pose (tilt angle $\alpha = 0$), parameter L becomes very sensitive to changes in α. Together with a thin CT slice thickness, this allows for precise placement of the needle within the CT scan plane.

The second image feature is the *deviation angle* $\Delta\phi$ between the actual and the desired needle orientation in the image. This parameter is used for the needle alignment feedback loop within the CT scan plane. Thus, the image feature error function can be defined as $\Delta f = f_d - f(L, \Delta\phi)$. This error function vanishes in case of (i) perfect needle location within the scan plane, and (ii) perfect needle alignment with the target. Because of the discrete representation of the needle pose in the image, a small interval around the desired image feature has to be admitted. Therefore, the stopping criterion of the visual servoing feedback loop has been defined as $|\Delta f| = |f_d - f| < \varepsilon$ with $\varepsilon = 0.2°$ (compare 6.3.3). If the needle is rotated within the scan plane, changes in the needle angle ϕ are directly visible. Thus, a determined deviation angle $\Delta\phi$ in the image can directly be used for robot control and leads to a proportional control law for this CT application. Ideally, only one iteration is needed to align the needle with the target[21].

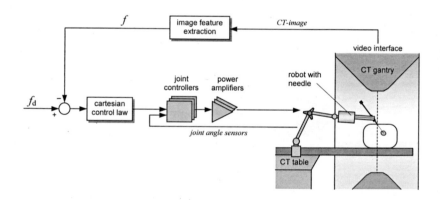

Figure 6.16: Controller block diagram of the visual servoing feedback loop for automatic needle alignment using CT-guidance.

---

[21] This requires the initial registration of the scan plane to the end-effector coordinate system of the robot, which is automatically done during the alignment procedure.

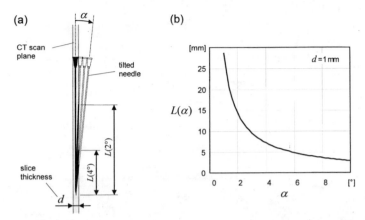

Figure 6.17:   (a) The visible needle length L in the CT image is an important parameter to control the automatic tilting of the needle into the CT scan plane. In case of precise placement of the needle within the scan plane ($\alpha = 0$) L increases to a maximum. (b) Close to this desired needle pose, parameter L becomes very sensitive to changes in the tilt angle α (see diagram). This is advantageous for precise positioning of the needle within the scan plane.

6.4.2 Workflow and User Interface

This section describes the workflow of the new approach for automatic CT-guided needle placement. An overview is given in figure 6.18. A special user interface has been implemented on the visual servoing workstation (see figure 6.19). Initial diagnostic scanning of the anatomical region of interest visualizes the target area. The intervention is planned and simulated on a workstation with the help of a reconstructed 3D volume of the scanned anatomy. Planning implies the definition of the *target point*, the most appropriate *access trajectory* and the resulting *insertion point*.

Depending on whether target and insertion point are located in the same scan plane or not, the ensuing procedure differs slightly, especially for the needle insertion itself (see scenario (a) and (b) on page 84). If the target and the insertion point are *located in the same scan plane*, it is obvious that the desired needle trajectory is lying in this plane as well. Therefore, the needle can be permanently observed in the CT images during the whole procedure of needle alignment and insertion. This allows visual monitoring by the operator as well as the automatic visual servoing control. Furthermore, needle drifting out of the image, which can lead to failure of the puncture, is directly recognizable in the image (compare figure 6.19).

The situation is quite different in case that target and insertion point are *located in different scan planes*. Here, the trajectory is oriented oblique to the CT scan plane and the image shows either the target or a cross section of the needle. In case of computer controlled and motor driven needle insertion, the needle tip can be automatically tracked by the visual servoing control, i.e. the CT table is moved horizontally by the visual servoing workstation in order to locate the needle tip permanently in the scan plane during automatic needle insertion. Beside the actual needle tip location in the image, an additional mark is drawn onto the CT image which symbolizes the desired tip position. The difference between the actual and the desired needle tip location gives the user information about the current amount of needle deviation (compare figure 6.15-c). This allows a kind of visual monitoring by the operator, even in case of needle insertion tilted to the image plane.

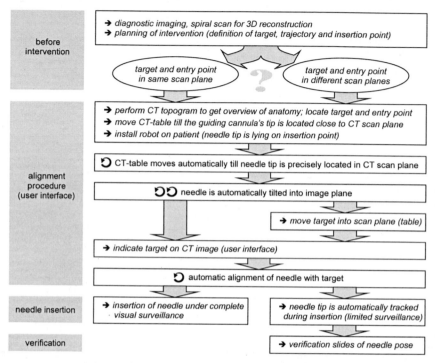

Figure 6.18: Workflow of new approach to automatic CT-guided needle alignment. The circular arrow symbolizes the visual servoing control loop (↻).

Before starting the intervention, a topogram of the region of interest is performed, showing a projective 2D view of the anatomy. This is quite similar to that of an X-ray radiograph. The topogram gives an overview of the anatomy and shows the approximate location of the scan planes containing the target and the insertion point. Then, the table is moved to locate the desired insertion point in the scan plane. Now, the manipulator has to be installed on the patient. For that purpose, the robot is tilting the guiding cannula so that only its tip is located in the scan plane at the insertion point. The needle tip shows up in the CT image as small bright spot (compare figure 6.20-1). Now, the user interface and control software on the visual servoing workstation have to be started. The layout of the user interface for CT-guided needle placement is shown in figure 6.19.

The first step in this procedure is the precise placement of the needle tip in the scan plane by automatic horizontal CT table motion[22]. This visually controlled positioning is started by pressing the corresponding control button on the user interface. Both workflows, needle insertion in the scan plane and oblique to the scan plane, are represented by a group of control buttons in the user interface (marked with (2) and (3) in figure 6.19). After this first control

---

[22] Actually, the guiding cannula's tip is placed in the scan plane. This is equivalent to placing the needle tip in the scan plane, if the needle is already put into the guiding cannula. In all these experiments the alignment procedure is performed with the needle placed in the guiding cannula, since this may lead to higher alignment precision (better control of parameter L).

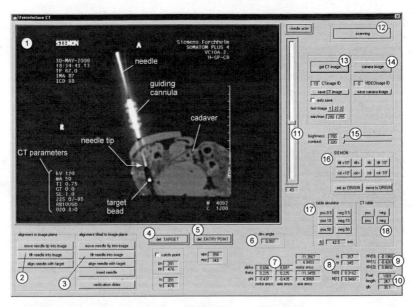

Figure 6.19: The user interface for automatic CT-guided needle placement.
(1) display of CT fluoroscopy image (video signal)
(2) control buttons for stepwise needle alignment in the image-plane
(3) control buttons for stepwise needle alignment tilted to image-plane
(4) button to confirm the target point in the image (image coordinates)
(5) button to confirm the entry point of the needle (image coordinates)
(6) deviation angle between needle axis and the desired orientation through target
(7) display of end-effector angles (Euler angles), robot axes and encoder angles
(8) display of needle rotation point and needle axis vector (image coordinates)
(9) display of real 3D needle vector in robot coordinate frame
(10) several parameters of the segmented needle in the current CT image
(11) slider for remotely controlled needle insertion
(12) button to activate/deactivate CT scanning
(13) button for acquisition and display the current CT image (from frame grabber)
(14) button for acquisition and display of CCD images (CCD-camera)
(15) sliders to adjust image brightness and contrast (parameters for frame grabber)
(16) simple control buttons for robot end-effector to tilt and rotate the needle
(17) simple control and of the table simulator; display of table simulator position
(18) manual incremental movement of the CT-table

step, only the guiding cannula's tip is visible in the image as a bright spot (compare figure 6.20-1).

Now, the operator may press the next control button for automatic tilting of the guiding cannula into the image. This is done by the robot for two times in order to obtain two needle poses for definition of the scan plane in robot coordinates (symbolized with ꙨꙨ in figure 6.18; compare figure 6.20-2,3). In case of needle insertion tilted to the scan plane, the operator has to move the CT table to locate the target in the scan plane (otherwise the target would be already visible in the image).

Now, the target point has to be indicated in the CT image by placing the target marker with the mouse (see figure 6.20-4). The coordinates are displayed by the user interface (4). In the ensuing step, the control computer automatically starts the visual servoing control loop for needle alignment with the target. A detailed description of the mathematical background of this approach has already been presented above. Now, the needle may be inserted manually or, if the robot is provided with a needle drive, remotely controlled by the operator with a slider on the user interface (11). The current insertion depth of the needle is simultaneously displayed. After the needle has reached the target, the operator can perform verification slides (3), in order to determine the needle deviation with regard to the target.

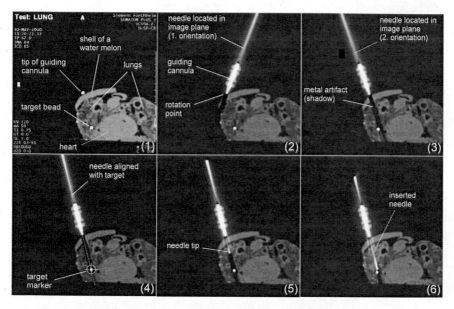

Figure 6.20:   CT-scans taken during the automatic needle placement procedure in case that insertion point and target are located in the same scan plane. A small metal bead
($\varnothing$ 2mm) serves as target which is placed in the lung of a pig cadaver. (1) The guiding cannula's tip is visible in the scan plane as a bright spot. The robot is automatically tilting the guiding cannula into the image for two different needle poses (2)(3).
Then, the needle is moved in the scan plane till alignment with the target is achieved
(4). Now, the needle may be inserted remote-controlled or manually (5)(6).

CT scanning is activated by the visual servoing workstation via foot pedal interface at the CT gantry. Scanning can be manually released by pressing the scan button on the user interface (12), or automatically by the computer during the needle alignment procedure.

The user interface provides several additional buttons and parameter displays for manual control of the robot and image acquisition. For example, the robot can be easily remote-controlled by several control buttons (16). A similar incremental control is provided for moving the table simulator (17) or the CT table (18). The user interface allows the acquisition of CT images (13) or images taken by a CCD camera (14). The CCD-camera is employed for deviation computation after needle alignment in preliminary CT experiments (see section 8.2.1). The actual status of all robot angles computed from the joint encoder signals and the DC motor encoder signals are displayed (7). Additional parameters computed and displayed during the alignment procedure help the user to monitor the procedure. Parameters are for example the needle rotation point in the image (rx, ry), the needle-axis vector (N_0, N_1) in the image (8), as well as the 3D needle-axis vector NV in robot coordinates (9) or additional information about the segmented needle in the image (10).

Figure 6.20 shows the workflow of the automatic needle placement procedure by means of CT-scans. The images show a piece of a pig cadaver (see section 7.4.3). The metal artifact (shadow) in the axis of the needle results from complete X-ray absorption along the needle [174]. In this series the insertion point and the target bead are located in the same scan plane.

6.5 Image Feature Extraction

The automatic image-guided needle alignment approaches presented above are based on processing visual information for control of the robot. This implies *image processing* and *image interpretation*. The former is the enhancement of images such that the resulting image more clearly depicts certain characteristics. The latter typically involves the extraction of a small number of numeric *features* from the image [35]. In case of the presented needle placement approach these image features are for example the *needle orientation* or the *target location* in the X-ray images.

In the needle alignment experiments imaging has been provided by X-ray fluoroscopes, by a single slice CT scanner, and - in preliminary experiments - by a CCD camera. However, images acquired by an X-ray fluoroscope or a CCD-camera are commonly geometrical distorted. There are different effects leading to image distortions. However, they impede precise geometrical measurements in the image and falsify image feature extraction for geometrical computations, as used in the needle placement experiments. The following section deals with this problem of image distortions and shows how these images are corrected for image feature extraction. Section 6.5.2 introduces the image features used for the visual control of the robot and discusses the methods employed to extract them out of the images.

6.5.1 Elimination of Image Distortions

X-ray fluoroscopes commonly use image intensifiers for acquisition of X-ray images. However, the use of image intensifiers produce radiographs which are spatially distorted. This is affecting the accuracy with which one can measure objects and distances in the image [135]. Several factors relating to the image formation process occurring within an image intensifier combine to produce an output image with spatial distortion. These factors result in two types of spatial warping: pincushion distortion and S-distortion [136]. Pincushion distortion results from the curved geometry of the X-ray detector, which is leading to radial warping in the image. The second type of distortion is caused by the deflection of the local magnetic field (basically the earth's magnetic field) on moving electrons within the image intensifier. Both effects are getting recognizable in the image the more the region of interest is away from the image center (compare figure 6.21).

However, the approach of automatic X-ray guided needle placement requires precise measurement of the needle position and orientation in the image to guarantee successful targeting. Furthermore, the cross-ratio based estimation of the target depth, presented in section 6.3.1, requires accurate determination of marker distances on the needle phantom (compare figure 7.7). Therefore, it is necessary to perform an *image distortion correction* procedure before determining distances in the image or extracting image features.

For the development and implementation of the X-ray guided needle placement control software, a CCD-camera has been used for image acquisition (see section 7.3.1). This allowed to test the principle accuracy of the image-guided alignment technique very conveniently, without having to care about X-ray protection and safety considerations. However, a CCD-camera causes image distortions as well. These so-called *lens distortions* in the image lead to similar problems for image feature extraction as with an image intensifier. Lens distortions are generated by the optical lens of the CCD camera. Again, this effect is getting worse the more the region of interest is away from the image center.

There exist several techniques for distortion correction. All of them use a calibration phantom, mostly a marker plate on the intensifier, which is creating a regular pattern in the image. This warped image pattern is used to measure the spatial distortions, required for the adjacent dewarping process.

First experiments in automatic image-guided needle placement have been performed by the author at the 'Computer Integrated Surgery Lab' (CIS-Lab), Johns Hopkins University, Baltimore, USA. The X-ray C-arm which has been used in these experiments produced highly distorted radiographs. Therefore, a special aluminum calibration plate has been employed for image dewarping, which was provided by the CIS-Lab [179]. This semi-radiolucent aluminum plate has been placed over the detector of the fluoroscopic C-arm. Horizontal and vertical grooves were machined in a square pattern into the plate (checkerboard). These grooves show up as pale lines on the X-ray image, but still providing enough contrast to be found by an image processing algorithm developed by the CIS-Lab (see figure 6.21). Before starting the needle placement experiments, the calibration of the C-arm has been performed for predefined C-arm positions providing certain X-ray views of the test object. After calibration of the X-ray C-arm in predefined positions, the aluminum plate has been removed again from the image intensifier before the needle placement experiments were started.

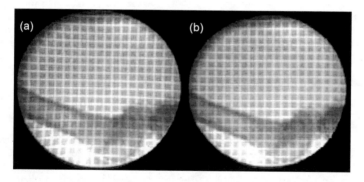

Figure 6.21:   Radiographs of the employed calibration plate used in an experiment performed by the CIS-Lab, Johns Hopkins University, Baltimore, USA [179]. (a) Fluoroscopic X-ray image of a bone. The white lines correspond to grooves cut into an aluminum dewarping calibration plate. (b) Dewarped image.

A very similar approach has been tested for distortion correction of the CCD-images, which have been acquired during the needle placement experiments. The lens distortion of the employed CCD-camera was tested with a checker pattern, printed on a sheet of paper with a laser printer (see figure 6.22). However, the CCD-images taken from the pattern showed very less distortions (figure 6.22-a), except at the image corners. A test showed that dewarping the image with the distortion correction algorithm provided by the CIS-Lab, did not perceptibly increase the quality of the images (see figure 6.22-b). Therefore, the CCD-images have been directly used for image feature extraction without preceding distortion correction. Furthermore, the image corners have been ignored for image feature extraction.

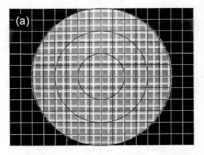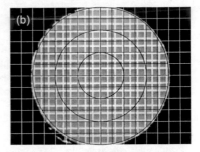

Figure 6.22:   Testing the grade of lens distortions in the CCD-images. We took a CCD-image of a
special checker pattern, printed on a sheet of paper, to determine the warping in the
image. The crossed lines projected onto the image allow a comparison of the pattern
with perfect horizontal and vertical lines. (a) The CCD-image (without distortion
correction) shows less distortion. (b) Distortion correction of the CCD-image does
not perceptibly increase the image quality.

In further X-ray fluoroscopy guided needle placement experiments, performed by the author
at the Siemens Medical Basic Research Department, Erlangen, Germany, a different type of
mobile X-ray C-arm system has been employed (Siremobil Iso-C, Siemens AG, Germany).
Again, the magnitude of image distortions had been determined before beginning with the
experiments. Distortions were tested with a special calibration plate fixed on the image intensifier. This calibration plate is provided with a regular pattern of small metal beads. Figure
6.23 shows a radiograph taken with this marker plate. The horizontal and vertical lines drawn
on the image demonstrate, that the distortions caused by the intensifier are negligible, even
near the image border. Therefore, the grabbed X-ray images are directly used for image processing without prior distortion correction.

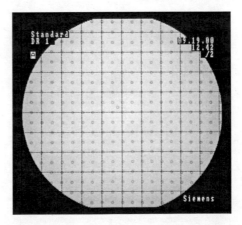

Figure 6.23:   Testing the grade of image distortions in the radiographs taken with the Siemens
Siremobil Iso-C system. The metal beads of a calibration marker plate, fixed on the
image intensifier, show up clearly in the radiograph. The horizontal and vertical
lines projected onto the image demonstrate the negligible distortions in the image.

6.5.2 Image Analysis – Extraction of Image Features

3D alignment of the robot's needle with a certain target requires needle alignment with the target in different image scenes. Figure 6.24-a shows an X-ray image taken during the needle alignment experiments performed at the Johns Hopkins Medical School with the RCM robot. In automatic X-ray- and CT-guided needle placement the basic parameter required to control the robot is the deviation angle $\Delta\phi$ between the actual and the desired needle orientation in the image. $\Delta\phi$ can be easily computed if the needle pose and the target location are visible in the image (compare figure 6.24-b). Therefore, the desired image features which have to be extracted from the digital radiographs are:

- needle orientation (needle vector n),
- needle tip (rotation point = entry point E),
- target location T (only in case of target tracking).

In computer vision the process of extracting e.g. the needle or the target from the image is commonly referred to as 'segmentation'. The problem of robustly segmenting a scene is of

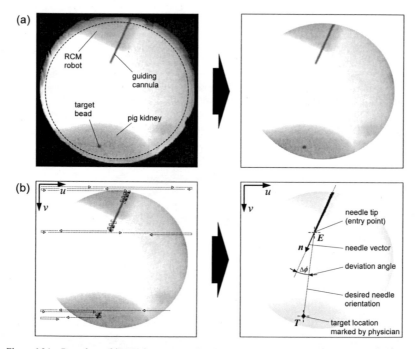

Figure 6.24:   Procedure of image feature extraction in case of X-ray guided needle placement. (a) First step: *image processing*. The black edge and the frayed image border of the original radiograph are removed. Now, the darkest regions in the image are the needle and the target bead. (b) Second step: *feature extraction*. The 'searching algorithm' starts from the top of the image and checks each pixel line-by-line whether the color value is smaller than a predefined threshold τ. If this is the case, the upper top of the needle is found in the image. The needle contour is followed down to the tip while extracting all joining pixels in the region. Then, the 'search algorithm' continues line-by-line in order to detect the region of the target bead. Finally, all desired image features can be computed: *needle vector, entry point, target location*.

high importance in computer vision. Much has been written about this topic and many methods have been described in literature [66][12]. But unfortunately many of the algorithms are iterative and time-consuming and thus not suitable for real-time applications [35]. In the automatic X-ray- or CT-guided needle placement experiments, a simple *threshold approach* has been implemented for the detection of the needle and the target bead in the radiographs.

The acquired images are represented by a pixel matrix with a size of 640×480 (NTSC video format), and an 8-bit grayscale color depth (0..255; 0=black, 255=white). Therefore, the grayscale image can be described as:

$$f(x,y) \in \{0,1,...,255\} \qquad \text{with} \qquad \begin{matrix} x = 0,1,...,639 \\ y = 0,1,...,479 \end{matrix} \qquad (6.8)$$

The process of image feature extraction is demonstrated in figure 6.24. In a first step, the black edge and the frayed image border of the original radiograph are removed. Then, the algorithm assumes that the needle and the target bead are the darkest regions in the image.

In the X-ray-guided experiments the needle is typically located at the top of the image while the target bead is placed somewhere below. A simple 'search algorithm' has been developed and implemented, which starts from the top of the image and checks each pixel line-by-line whether the color value $f(x,y)$ is smaller than a predefined threshold $\tau \in \{0...255\}$ (compare figure 6.24-b) [23]. If this is the case, the upper end of the needle in the image has been found. The 'search algorithm' follows the needle contour down to its tip and extracts all joining pixels of the region for subsequent computation of the needle axis.

In case that the target bead shall be extracted as well, this 'search' process can be continued for the proximate image lines in order to detect the pixels of the target. However, the workflow in both needle placement applications intends the *target to be defined manually* by the operator via user interface (compare section 6.3.4 and 6.4.2). Therefore, an automatic target detection has not been implemented.

As soon as the segmentation process has been finished, the needle axis can be computed out of all extracted needle pixels. This has been done by a simple line fit approach (compare figure 6.25). A very appropriate method for fitting a straight line to a set of points is the *Eigenvector line fitting* method [85]. In particular, a line will be the best fit to a set of points if it minimizes the sum of squares of the perpendicular distances Σx_i^2 from each point to the line. In contrast to common *least square methods*, the result is independent of the choice of the image coordinate axes u and v. Using a least square approach, the sum of squares, $\Sigma \Delta u_i^2$ and $\Sigma \Delta v_i^2$, are minimized in direction of the image coordinate axes. This may cause accuracy problems for nearly horizontal or vertical needle orientations. However, this is often the case in the presented needle placement experiments, thus the Eigenvector line fitting method is the favored approach to determine the needle axis in the image.

The needle in the image is represented by a set of pixels $p_i = (u_i, v_i)$, $i = 1..m$, in coordinate frame $[u_s, v_s]$. The origin of this coordinate frame is located at the center of gravity S of the needle pixels. This simplifies the computation of the desired needle axis, since the best line passes through the origin of this coordinate frame $[u_s, v_s]$. The desired vector n, representing the needle axis, will be derived by minimizing the distances Σx_i^2 from the needle pixels to the

---

[23] This approach is equivalent to an initial transformation of the 8-bit grayscale image $f(x,y)$ to a binary image $f^*(x,y) = \begin{cases} 0; & f(x,y) \le \tau \\ 1; & f(x,y) > \tau \end{cases}$

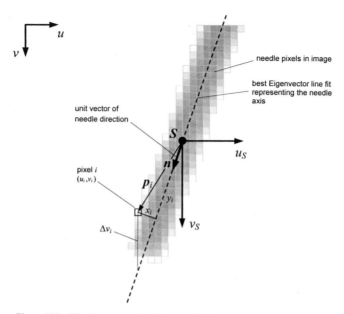

Figure 6.25: The *Eigenvector line fitting* method is used to determine the needle axis in the image. This approach minimizes the sum of squares of the perpendicular distances Σx_i^2
from each point p to the line. In contrast to common *least square methods*, where
the sum of squares, $\Sigma \Delta u_i^2$ and $\Sigma \Delta v_i^2$, are minimized in direction of the image coordinate axes u and v, the Eigenvector line fit is independent of the choice of the coordinate frame. S is the center of gravity of the region of needle pixels.

needle axis. This is equivalent to maximizing the sum of squares of their projections Σy_i^2 onto
the needle axis:

$$\Sigma y_i^2 = \Sigma \left(n^T p_i \right)^2 = \max$$
$$= n^T \Sigma p_i p_i^T n$$
$$= n^T \cdot A \cdot n \qquad with \quad A = \begin{bmatrix} \Sigma p_{1i}^2 & \Sigma p_{1i} p_{2i} \\ \Sigma p_{1i} p_{2i} & \Sigma p_{2i}^2 \end{bmatrix} \qquad (6.9)$$

The symmetric matrix A is the scatter matrix of the needle pixels p_i in the image and is, except the division with $(m-1)$, similar to the variance and covariance of these points. The best
line can be derived with an Eigenvalue approach using the characteristic equation of A as
follows:

$$(A - \lambda I) n = 0. \qquad (6.10)$$

This equation can be solved by calculation of the determinant $|A - \lambda I| = 0$, which is leading to
the Eigenvalues

$$\lambda_{1,2} = \frac{1}{2}\left((A_{11}A_{22}) \pm \sqrt{1 - 4(A_{11}A_{22} - A_{12}^2)}\right) \tag{6.11}$$

The Eigenvector associated with the largest Eigenvalue has the same direction as the best fitting line and is therefore representing the needle axis with respect to the image coordinate frame $[u_S, v_S]$. Since the Eigenvector n of the square matrix A is the solution vector that satisfies equation (6.10), the desired Eigenvector n is derived by substituting the largest Eigenvalue λ^* in (6.10) and subsequent solving for n:

$$n = \begin{bmatrix} 1 \\ A_{12}/(\lambda^* - A_{22}) \end{bmatrix} \quad \text{with} \quad \lambda^* = \max(\lambda_1, \lambda_2). \tag{6.12}$$

After normalization of the needle vector n, the direction is inverted if n does not point towards the target. So finally, the implemented Eigenvector line fitting method results in a needle vector which is pointing in needle insertion direction towards the target (compare figure 6.24 and 6.25). The projected needle length L in the image is derived from the largest distance of a needle pixel p_i to the center of gravity S.

$$\frac{L}{2} = \max|p_i|. \tag{6.13}$$

Thus, the skin entry point E (needle tip) can be computed in image coordinate frame $[u,v]$ as

$$E = \max|p_i| \cdot n + S. \tag{6.14}$$

Now, the deviation angle $\Delta\phi$ between the actual needle axis n and the desired needle orientation $\overrightarrow{ET}$ (compare figure 6.24-b) can be easily computed via vector product:

$$\Delta\phi = \arcsin \frac{\left|n \times \overrightarrow{ET}\right|}{|n|\left|\overrightarrow{ET}\right|}. \tag{6.15}$$

This deviation angle $\Delta\phi$ is a basic parameter required to control the needle-guiding robot. Furthermore, in automatic CT-guided needle placement the needle length L in the image is needed to control the automatic tilting of the needle into the CT scan plane (compare section 6.4.1).

A further challenge in image feature extraction, which has not been mentioned yet, is demonstrated in figure 6.26. The zoom-in of the guiding cannula (a) and the target bead (b) demonstrate a basic problem of image feature extraction: the discrete and blurred edge, which makes a precise determination of the object boundary - and hence the object size - difficult. In this context the appropriate choice for threshold τ is of high importance for reliable segmentation and feature extraction. Accordingly, threshold τ has been adjusted during the experiments for appropriate results.

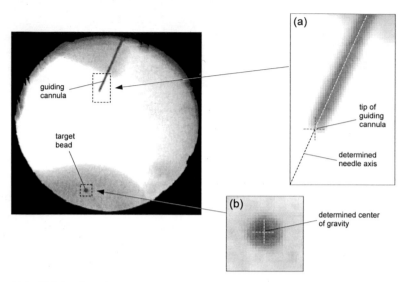

Figure 6.26: Digital radiograph taken during the needle alignment experiments with a pig kidney. The
zoom-in of the guiding cannula (a) and the target bead (b) demonstrate the discrete and blurred
edges, which makes a precise determination of the object boundary (size) difficult.

6.6   The Visual Servoing Workstation

A basic component in both the X-ray- and the CT-scenario is the *visual servoing workstation*,
which integrates all hardware and software components for robot control, image acquisition
and image analysis. Figure 6.27 shows a schematic of the system architecture in case of employing an X-ray C-arm. Although, the hardware configuration of the visual servoing workstation is the same for both imaging modalities, the control software is slightly different. This
is because of the mathematically different needle alignment approaches in both scenarios, and
due to the analysis of different image types (X-ray radiographs versus CT slices). The basic
hardware and software component of both setups are presented in the following paragraphs.

Whatever imaging modality is used for automatic visually controlled needle placement, X-
ray fluoroscopy or CT-imaging, the visual servoing workstation has to gain basic control
functionality over the imaging system, for example image release. However, getting sophisticated control over an X-ray C-arm or a CT scanner is very challenging, since these systems
commonly do not provide any interface for the connection of external systems which may
take the control over the imaging modality. Therefore, the control functionality of the visual
servoing workstation over the imaging modality is very restricted and limited basically to image release. The technical realization for both modalities is described below.

The system employed for the automatic, visually controlled needle placement experiments
comprises several components integrated in the visual servoing workstation (as plug-in interface boards for low-level communication with the manipulator and the imaging system). The
following sections present the hardware and software architecture of the visual servoing
workstation.

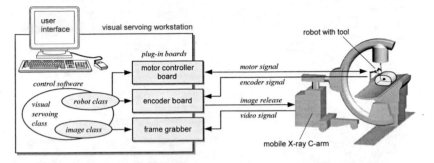

Figure 6.27: Schematic representation of the system design for automatic, visually controlled needle placement (here with an X-ray C-arm). The visual servoing workstation integrates all hardware and software components for robot control and the image acquisition: *hardware*, including the interface boards for the low-level communication with the manipulator and the imaging system; *software*, including the user interface and the visual servoing control software.

6.6.1 Hardware Components

The visual servoing workstation integrates several plug-in boards for robot control, image acquisition, and for basic control functions of the X-ray C-arm and the CT-scanner. The workstation itself comprises of a Pentium 200MHz desktop PC with the Windows™ NT operating system. Figure 6.28 shows the workstation with its various interface boards and connectors.

The needle-guiding robot is controlled by a 2-axis DSP-based controller board (C-842, Polytec PI, Inc., Tustin, CA, USA). This ISA plug-in board is based on a fast DSP processor providing high performance PID motion control using the MC1201 Advanced Multi-axis Motion Control chipset. The board is designed as a 16-bit ISA bus board. It provides closed-loop digital servo control using incremental encoder position feedback signals. The axes can be programmed in synchrony to allow advanced multi-axis motion. The C-842 is controlled by a host operating program which interfaces with the board bi-directional via I/O ports. The motors can be controlled in position, velocity, acceleration and other motion relevant properties. The board provides limit switch control, and 8 TTL input and 8 TTL output channels [183]. An NT-driver package is provided including a software DLL to accomplish board access. Function libraries in C++ offer sophisticated and direct access to the C-842. An extended command set has been implemented into the visual servoing software (*robot class*).

The PC also houses a pulse acquisition card (PC-Puls, MEGATRON Elektronic, Putzbrunn, Germany) for the incremental axis encoders inside the robot. This ISA plug-in encoder board provides 3 inputs for TTL signals (encoder) and 4 programmable relay outputs. These relay outputs are used for basic control of the CT scanner and the X-ray C-arm, e.g. image release or incremental CT table movement. A driver DLL for Windows NT and a software library in C allow easy control and implementation in the visual servoing software (*robot class*).

To operate the stepper motor of the CT-table simulator, a power electronics unit for a 5-phase stepping motor has been integrated into the workstation (D450.00, Berger-Lahr GmbH, Lahr, Germany). This power unit provides a digital signal input interface (TTL signal input level) for setting of the direction of rotation, the stepping mode (up to 1000 steps per revolution), and the pulse input. These TTL input signals are generated by the 8 programmable TTL

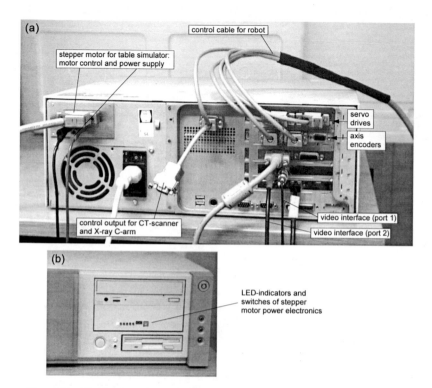

Figure 6.28: The visual servoing workstation (Pentium 200MHz PC) housing all hardware for robot control and image acquisition. (a) The back view shows the various connectors for control of the robot, the CT table simulator, the CT-scanner (or X-ray C-arm), and for video image acquisition. (a) Several LED-indicators and switches for the stepper motor power electronics are easily accessible from the front side.

output channels of the employed C-842 motor controller board. LED-indicators in the device housing show the status of the power unit (see figure 6.28-b). The visual control software (*robot class*) provides several commands to control the stepper motor power electronics unit via motor controller board.

Image acquisition is performed by a frame grabber (FALCON, IDS Imaging Development Systems, Obersulm, Germany) which is integrated in the visual servoing workstation. This PCI plug-in board supports video standards like NTSC or PAL with a maximum resolution of 768×576 pixels. The FALCON frame grabber supports up to two cameras and allows for various settings, like image size, contrast, brightness or the color format (e.g. RGB32 or 8bit gray scale). The video images are digitized and transferred into the PC RAM for image processing. A hardware driver for Windows NT and a software development kit allowed easy implementation into the visual servoing software (*image class*).

6.6.2 Software Components

The principle and theory of the novel visual servoing approach for X-ray guided and CT-guided needle placement has been presented in section 6.3 and 6.4. The implementation of this image based control is realized in an object-oriented C++ software library, programmed with the MS Visual C++ compiler. Figure 6.29 depicts the basic structure of the visual servoing control software. It consists of two basic software modules which has been developed for hardware control and communication with the manipulator: (i) the *robot class* for controlling the manipulator, and (ii) the *image class*, which provides various functions for image acquisition and processing. The visual servoing control loop is embedded in the *visual servoing class* (parent class), which makes use of both the *robot class* and the *image class*.

The *robot class* provides the communication and control of the robot drives and the robot joint encoders. It integrates the hardware drivers for the motor controller board and the encoder card. Table 12.1 (Appendix) gives an overview of the member functions of the robot class. After initializing the motor controller and the encoder board, several functions for absolute and relative motion control of the robot axes and needle drive are provided. Evaluation of the axes encoder signals allows for very precise needle positioning. The stepper motor of the CT table simulator is driven by the motor controller board as well. The power electronics unit of the stepper motor is receiving TTL signals by the programmable TTL output channels of the motor controller board. Additionally, the robot class controls the programmable relay

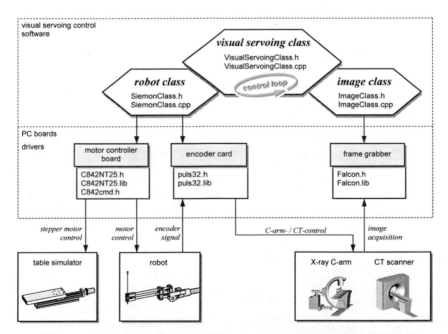

Figure 6.29: Basic structure of the visual servoing control software and the interface to the hardware. The visual servoing control loop is integrated in the *visual servoing class*, which includes the *robot class* for controlling the manipulator, and the *image class* for image acquisition. The employed software drivers (H-files and LIB-files) are listed below each PC board.

outputs of the encoder card to support basic control functions of the CT scanner or the mobile X-ray C-arm (e.g. image acquisition release or incremental CT table motion). A brief description of these functions is given in the appendix in table 12.1.

The *image class* is used for image acquisition and allows for communication and control of the frame grabber board. It integrates the frame grabber hardware driver and provides several software functions for image acquisition and processing (see table 12.2, Appendix). For example, all algorithms for image feature extraction presented in section 6.5 have been implemented in several functions of the *image class*.

Another fundamental software component is the user interface, which has already been presented in section 6.3.4 and 6.4.2 for the particular imaging modality. Although, the user interface is essential in both setups, it has no impact on the visual servoing approach and is therefore not further described here.

Chapter 7
Needle Placement Experiments Using X-ray Imaging

This chapter is presenting the experiments conducted in automatic X-ray-guided needle placement. It is divided in two main parts: (i) the experimental setup with the implementation of all appending components, and (ii) the presentation of the experiments for evaluation of the novel approach in automatic X-ray-guided needle placement. In each experiment imaging is provided by a mobile X-ray C-arm.

7.1 Background

By the end of 1998, the author has been in the situation to have a novel and comprehensive theoretical framework for automatic X-ray-guided needle placement, but without any manipulator (robot) for experimental evaluation of this new approach. At the same time the Siemens AG started a collaboration with the CIS Lab (Computer Integrated Surgery Lab), which is part of the Engineering Research Center of Computer Integrated Surgery, led by Prof. Russell H. Taylor at the Johns Hopkins University, Baltimore, USA. This group itself is working on robot-supported surgical interventions and appreciated the idea of using visual servoing for automatic needle alignment. Therefore, in 1999 the author conducted first experiments in automatic visually controlled needle placement at the Johns Hopkins University, supported by the CIS-Lab with different medical robots. In preliminary experiments a CCD-camera for image acquisition has been employed, which allowed to test the principle accuracy of the novel approach very conveniently without having to deal with X-rays and safety considerations. Ensuing experiments have been performed by the author at the Johns Hopkins Medical School, where the CIS-Lab could provide an uniplanar mobile X-ray C-arm for a second, more clinical experimental setup.

Back in Germany, the author started the development of the novel, miniaturized robot, which has been presented in chapter 5. This prototype was then used in all ensuing experiments in X-ray- and CT-guided needle placement at the Siemens Medical Solutions laboratories in Erlangen and Forchheim, Germany.

7.2 System Design for X-ray-guided Needle Placement

Commonly a mobile X-ray fluoroscope consists of the following components (see figure 7.1): (i) The mobile X-ray C-arm: its C-shaped arm with the X-ray tube on the one side and the image intensifier (detector) on the opposite side, can be moved and rotated around two axes in order to provide various X-ray views. The entire C-arm control and X-ray generator unit is integrated in the chassis of the mobile system. (ii) The monitor rack with the digital imaging system, a video recorder, and a video printer. The C-arm operating system in the chassis controls X-ray image acquisition and sends the digital image raw data to the imaging system. There, the raw data is processed and images are displayed on the monitors. Additionally, the actually acquired image is provided as NTSC video signal by a video output of the imaging system. For documentation all images are stored on an archiving medium.

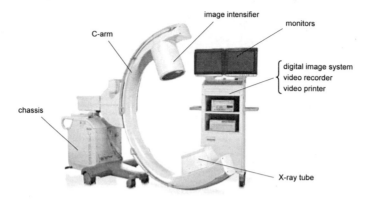

Figure 7.1: Typical configuration of a mobile X-ray C-arm (SIREMOBIL Iso-C, Siemens AG). The same type of X-ray fluoroscope has been employed for C-arm guided needle placement experiments at the Siemens Medical Solutions laboratories.

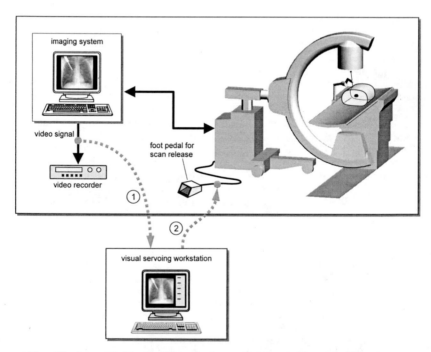

Figure 7.2: Setup of the X-ray-guided needle placement experiments. The acquired X-ray images are sent to the imaging system and displayed on the monitor. A video output provides the actual X-ray image as NTSC video signal. The operator can start and stop X-ray fluoroscopy via foot pedal. The visual servoing workstation takes control of the fluoroscopy system by simulating the foot pedal function (2) and acquires the X-ray image by the video output of the fluoroscope via frame grabber (1).

Additionally, the system comprises of a foot pedal for scan release (compare figure 7.2), with which the operator can start and stop X-ray fluoroscopy. The foot pedal integrates two switches. If the pedal is pressed, these switches close an electric circuit and indicate the operating system to start imaging. The foot pedal interface allows simple but convenient scan control and is used by the visual servoing workstation for remotely controlled scan release. Figure 7.2 shows a schematic of the X-ray fluoroscopy setup together with the connected visual servoing workstation. It demonstrates the interfaces and communication between the fluoroscopy system and the visual servoing workstation.

As described in section 6.6.1, the pulse acquisition board inside the visual servoing workstation provides four programmable relay outputs. Three of them are internally connected with the control output of the workstation (compare figure 6.28). In the visual servoing setup the foot pedal is disconnected and replaced by a cable, which connects the X-ray fluoroscope with the control output of the visual servoing workstation. Thus, scan release can now be easily performed remotely controlled by the visual servoing workstation instead of pressing the foot pedal.

For image acquisition the visual servoing workstation is connected to the imaging system of the X-ray fluoroscope (see figure 7.2). The images are provided via video interface and are simply captured by the frame grabber board of the visual servoing workstation.

7.3 Experiments at the Johns Hopkins University

In 1999, the author conducted first experiments in X-ray fluoroscopy guided needle placement at the Johns Hopkins University with three different medical robots provided by the CIS Lab: *LARS* (IBM research), *RCM* (Johns Hopkins University), and *Neuromate* (Integrated Surgical Systems, Inc.), which have already been presented in section 2.2.2 and 2.3.4. These three robots have different characteristics in size, kinematics, work envelope and dynamics. LARS and the smaller RCM robot have a fixed remote center of motion, defined by their parallel kinematic, while Neuromate has an articulated arm with 5 axes (see figure 7.3).

The LARS robot was developed jointly by the Johns Hopkins University and IBM Research to aid surgeons in laparoscopic applications including camera holding and precise instrument control for active assistance during laparoscopic procedures [158]. LARS is a 7 degree-of-freedom manipulator. It has three translation stages at the base, two rotational degrees of freedom at the shoulder, and two stages for instrument insertion and rotation. The robot has a stage resolution of 0.05 mm, and an overall positional accuracy close to 0.1 mm. Its parallel link construction rotates the instrument around a fixed rotation point (remote center of motion), which is identical with the insertion point during surgery. Sensors mounted on the in-

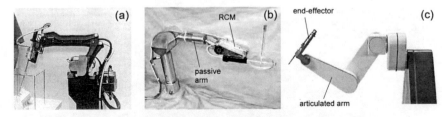

Figure 7.3: The three medical robots provided by the CIS Lab, which have been employed to perform first experiments in automatic image-guided needle placement: (a) *LARS*, IBM research (compare figure 3.8), (b) *RCM*, Johns Hopkins University, (c) *Neuromate*, Integrated Surgical Systems Inc.

strument carrier limit the amount of force and torque extended on the surgical instruments [127]. Should forces or torques exceed safety thresholds, the robot ceases all motion until they are again within safe limits or the operator intervenes. For the presented experiments, the instrument carrier was provided with a simple needle phantom.

A much smaller and more compact robot for needle placement is the RCM robot developed by the Johns Hopkins University [154]. It is a kind of end-effector which can be mounted on a passive or active arm (see figure 7.3-b). Again, the parallel kinematic provides two rotations of the needle around a fixed rotation point (Remote Center of Motion), comparable to the LARS robot. The compact design has just 1.6 kg in weight and may be folded into a 171×69×52 mm cube, but it has no force or torque sensors integrated as the LARS robot.

Neuromate is a sophisticated robotic system for stereotactic brain surgery (Integrated Surgical Systems, Inc., USA) [181]. The system consists of a 5-dof articulated robotic arm and a PC based kinematic positioning software system, for orienting and positioning of a surgical tool. Neuromate is the only one of the three employed robots, which is commercially available and which has the FDA approval for the US market.

Beside the manipulators itself, the CIS Lab provided an extensive software library for robot control. This C++ software library, called the JHU Modular Robot Control (MRC) [86], provides the machine level robot control functionality for each of their robot systems. It includes software classes for kinematics, joint level control, sensor support, peripheral support, and network support. Furthermore, a variety of I/O devices including serial and parallel ports, ATI force sensors, joysticks, digital buttons and foot pedals are supported [160]. The use of this MRC library allowed the fast implementation of various robots into the novel visual servoing application. The application using the MRC library does not need to know about the details of the hardware. Several robots can be controlled by the same application without changes in the application source code [86]. This is realized by a client/server network architecture, which is demonstrated in figure 7.4. The host computer with the visual servoing application is communicating with the robot server over the (local) network to control the robot.

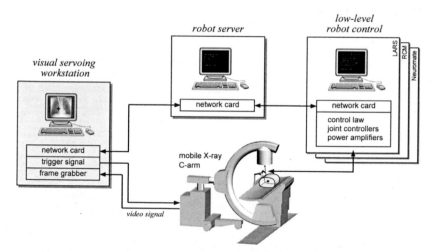

Figure 7.4: Diagram of the client/server network architecture for robot control. A software library (MRC), developed by the CIS Lab, provides the low-level robot control functionality for each of their robot systems. This simplifies the communication between the visual servoing application and the robot.

Because most robot motions are very slow, delays caused by the network had no bad influence on the control performance, since network distances are small. The host computer is logged on the robot server as a client. After the host computer declares the robot to be used to the server, the server automatically configures the robot control computer, which is responsible for the low level robot control functionality. Therefore, the motion control of the robot is limited to simple kinematic commands. The host does not have to know any hardware related information about the robot or its control. This allowed a fast and convenient implementation of the different robots into the visual servoing test bed.

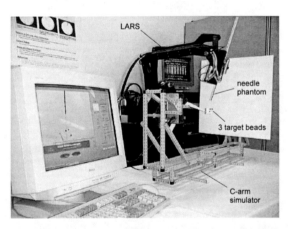

Figure 7.5: Experimental setup with CCD-camera for image acquisition. The LARS robot together with a special C-arm-simulator has been employed in the first experiments using a CCD-camera. The image processing computer with the user-interface is shown on the left side of the image.

7.3.1 Preliminary Experiments with CCD-imaging

Preliminary experiments have been performed at the CIS Lab using a CCD-camera for image acquisition. This allowed to test the principle accuracy of the image-guided alignment technique very conveniently without having to deal with X-rays and safety considerations. The experimental setup with the LARS robot is shown in figure 7.5. The host computer with integrated user-interface and visual servoing control is shown on the left side of the image. The computer with the low level robot control shows up in the corner behind LARS. In this CCD-setup a black needle phantom with four marker beads is used instead of a real needle. The markers on the needle are used to verify the depth estimation approach after needle alignment. The needle phantom is fixed to the robot's end-effector. A target holder with three beads is mounted to the base of the robot.

For holding the CCD-camera a special *X-ray C-arm-simulator* has been built, which is positioned on a table in front of the robot. It is a lightweight assembly with two vertical arms. The left arm is holding the CCD-camera, while a white cardboard is fixed to the opposite side (see figure 7.6). This is corresponding to an 'X-ray C-arm' with its X-ray tube and the image intensifier on opposite sides. Actually, the CCD-images acquired by the C-arm-simulator provided quite similar images compared to real X-ray radiographs of the needle and the target beads. Figure 7.7 shows a CCD-image taken during an alignment experiment with the C-arm

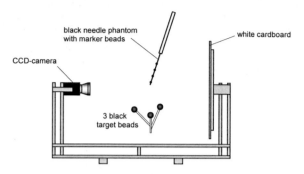

Figure 7.6: The C-arm-simulator: it consists of a lightweight assembly with a CCD-camera and a white cardboard on opposite sides. This provides us with quite similar images, both in the CCD- and the X-ray scenario.

simulator. The black needle and the three black target beads in front of the white cardboard (highlighted background) provide a CCD-image very similar to that of an X-ray fluoroscope (without contrasty anatomical structures in the background; compare figure 7.10).

Needle detection and segmentation is performed with a simple *threshold approach* in order to extract the desired image features (compare section 6.5.2). During the experiments it was intended to arrange repeatable conditions in the images, e.g. illumination of the scene, or the highlighted background. Therefore, the black needle generally showed up very clearly in the images and could be easily detected. As already described in section 6.5.2 the needle was commonly located in the upper half of the radiograph, which simplified the search for the needle location in the image. Both circumstances facilitated the detection of the needle sig-

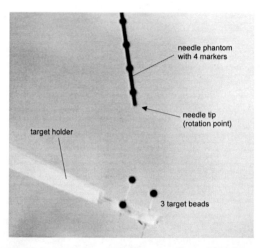

Figure 7.7: CCD-image of the needle phantom and three target beads taken with the C-arm-simulator. With a bright image background these CCD-images are very similar to that of an X-ray radiograph without contrasty anatomical background. For these experiments a special needle phantom is used, fitted with 4 marker beads with defined distances.

nificantly, and allowed to use a simple threshold algorithm, where the needle pixels are identified as grayscale values higher than a certain threshold τ, which can be adapted by the user interface (compare figure 6.9). For automatic estimation of the required insertion depth, the markers on the needle have to be localized automatically. To find the centroids of these marker beads, a special algorithm has been developed that follows the needle contour and identifies a marker bead as a rapid increase of the needle's width. As discussed in section 6.5.1, images acquired with the CCD-camera showed only very less distortions, especially in the region of interest. Therefore, the images have been directly used for image feature extraction without prior distortion correction.

The programming of the user interface software and the visual servoing control was done in the environment of the CIS Lab using the LARS robot and the C-arm-simulator. Two software libraries for robot control and image acquisition have been implemented, both developed by the Computer Integrated Surgery Lab: (a) the MRC robot control library, and (b) a software library for image acquisition via frame grabber. After successful CCD-experiments with the LARS robot, the robot control has been extended for two additional robots, the RCM robot and the Neuromate. Figure 7.8 shows both robots during needle alignment experiments with the C-arm-simulator. Due to the MRC robot control library, the control to move the end-effector (needle) e.g. around a certain point is identical for each type of robot. Therefore no significant changes in the source code have been required for the implementation of RCM or Neuromate.

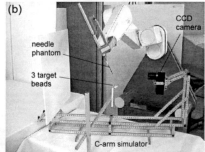

Figure 7.8: Additional experiments using CCD-imaging. (a) Set-up with RCM robot and 5 target beads. The parallel kinematic of the manipulator is clearly visible. In these experiments a real needle is used. (b) Setup with Neuromate together with the C-arm-simulator.

7.3.2 Needle Placement Experiments with an X-ray C-arm (JHU)

Initial verification of the novel approach using X-ray imaging was performed at the Johns Hopkins Medical School, where the CIS-Lab could provide a uniplanar mobile X-ray fluoroscope (GE Polarix 2). The C-arm had an approximate source-to-intensifier distance of 1 meter, and an intensifier diameter of about 14 cm. Figure 7.9 shows the first X-ray setup with the LARS robot and the fluoroscope.

The target has been a freestanding small metal bead with a diameter of 2 millimeters, which was clearly visible in the radiographs. For this setup the same needle phantom is used as in the CCD-experiments presented above. The diameter of 14 cm of the image intensifier was rather small, therefore the intensifier had to be placed as near as possible to the targeting volume.

In contrast to CCD-imaging, these X-ray experiments had to follow special safety rules to avoid radiation exposure for all involved persons. During the experiments everybody in the laboratory had to wear lead aprons for radiation protection. Furthermore, lead covered mobile walls are used to stand behind during radiation.

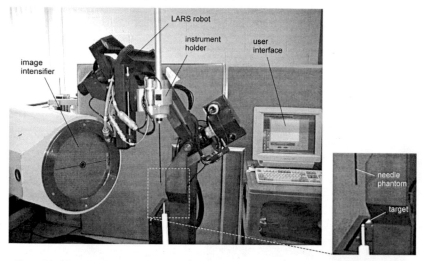

Figure 7.9: First experiments in X-ray guided needle placement at the Johns Hopkins Medical School. The LARS robot automatically performed image-guided needle alignment with a small metal bead (target). X-ray imaging is provided by a mobile X-ray fluoroscope (C-arm).

The images produced by the employed X-ray fluoroscope have been spatially distorted (compare section 6.5.1). Before they could be used for the extraction of geometric image features, an image distortion correction had to be performed. For this purpose the CIS Lab provided a special aluminum calibration plate, attachable in front of the C-arm detector, and the appropriate image processing software. The process of image distortion correction has already been described in section 6.5.1.

In a second X-ray setup the smaller RCM robot was employed for needle manipulation. Again the Ø2mm metal bead served as target, which has been positioned about 70 mm below the needle guiding cannula of the RCM end-effector. Figure 7.10 shows an X-ray radiograph taken after the automatic needle alignment process with the RCM robot. The inserted needle shows accurate alignment with the target in this view. During the alignment process only the metallic guiding cannula at the end-effector of the RCM robot has been visible, while its tip is identical with the rotation point. The frayed border of the radiograph is caused by the distortion correction process. The alignment procedure is identical to that described above for CCD-imaging, except that the black edge and the frayed image border are faded out before extracting the pose of the guiding cannula in the image.

After successful implementation and testing of the automatic image-guided needle alignment approach for X-ray fluoroscopy, a more clinical application has been chosen: the puncture of a pig-kidney under X-ray imaging. After organizing three fresh pig kidneys from a local slaughterhouse, the initial X-ray setup has been slightly modified as shown in

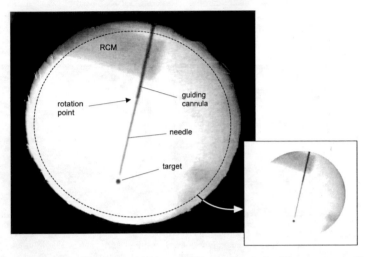

Figure 7.10: Example of an X-ray radiograph taken after the automatic needle alignment process with the RCM robot. The inserted needle shows accurate alignment with the target in this view. During the automatic alignment process only the metallic guiding cannula at the end of the RCM robot are visible. Before extracting the pose of the guiding cannula in the image, the black edge and the frayed image border are removed.

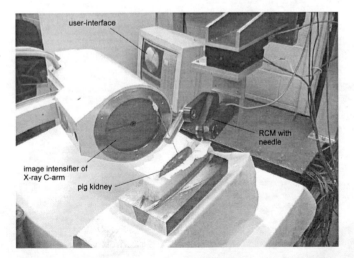

Figure 7.11: Puncture of a single pig kidney at the Johns Hopkins Medical School. In this setup the RCM robot with the mobile X-ray fluoroscope are employed. The automatic needle placement procedure is performed by the RCM robot, which is fixed to a three-axis portal robot. The needle is hold by a radiolucent acrylic needle holder.

figure 7.11. The kidney has been fixed in vertical position between foamed plastic, about 6 cm below the tip of the guiding cannula. The compact RCM robot allows closer positioning of the image intensifier to the robot and the targeting volume, compared to LARS. This is advantageous because of the narrow diameter of the image intensifier.

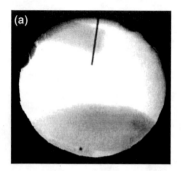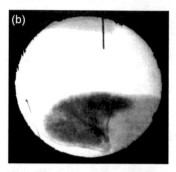

Figure 7.12:   X-ray images of a pig kidney. (a) A metal bead is implanted and serves as target. Without contrast medium the anatomical structures of the kidney are not visible. (b) After injection of contrast medium into one renal artery, the blood vessel shows up clearly.

Soft-tissues in a radiograph, like e.g. a kidney, show low contrast and are only hard to identify. After first experiments with an implanted metal bead as target, which showed up clearly and sharply bounded in the image (figure 7.12-a), the author decided for more realistic conditions: the puncture of a contrasted pig kidney. A catheter has been fixed to one arteria renalis interlobares in order to inject a contrast medium with a syringe (Hexabrix, Mallinckrodt Medical, St. Louis, USA). The purpose of applying contrast medium is to increase the density of the artery in relation to the surrounding tissues, so that its anatomical structure become visible in the X-ray image. But with time, the contrast medium diffuses into the surrounding tissues and the contours are getting increasingly blurred (compare figure 7.12-b). After con-

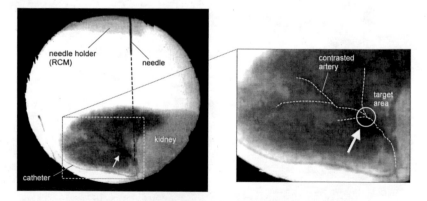

Figure 7.13:   Radiograph taken during the needle alignment experiments in the X-ray set-up with a pig kidney (see figure 7.11). Contrast medium is inserted with a catheter into the left renal artery. The first main ramification (see white arrow) will serve as target, since this point is clearly visible on the radiograph.

trasting the renal artery, the author decided to puncture the first main ramification of the vascular contour, since this point was clearly visible on the radiograph (see figure 7.13).

To evaluate the efficiency of this approach, obviously the accuracy with which the needle can be aligned to the target is an important factor. However, figure 7.13 illustrates a problem for the precise determination of the remaining deviation after needle alignment. While the center of a round metal bead is always clearly identifiable in each X-ray view (compare figure 7.12-a), the situation in targeting real anatomical structures can be much more challenging. The anatomical target might not be clearly visible from different viewpoints and, moreover, will change its shape. This is a problem not only for the physician in defining the target in the first and second X-ray view, but also for verification of the remaining needle deviation in the presented experiments. Obviously, it is hard to mark precisely two corresponding points from different viewpoints, if the target area is blurred and contours are not clearly recognizable (see figure 7.13). Therefore, the precise determination of the 'principle' accuracy of the novel approach is determined by targeting well defined metal beads. However, the accuracy which is achievable in real clinical routine can finally be investigated only by real test series on patients. The problem in defining the accuracy of the novel needle placement approach is extensively discussed in section 9.1.

7.4   Experiments at the Siemens Medical Solutions Laboratories

After four month, the experiments at the Johns Hopkins University had been completed. Motivated by the promising results, the author decided to develop himself a robotic manipulator, optimized for needle placement applications under X-ray and CT guidance. Based on the information obtained by the physician interviews (see chapter 4) and the experiences made at the Johns Hopkins University, the author defined a list of requirements and started to build up the manipulator presented in chapter 5. This prototype was employed in all ensuing experiments in automatic needle placement with X-ray or CT-imaging.

Initial evaluation of the prototype was performed in a laboratory at the Basic Research Department under controlled conditions. Basically, the new test bed and the conducted experiments in X-ray guided needle placement are similar to those at the Johns Hopkins University. The implemented visual servoing control software with its algorithms for image analysis and robot control is identical. However, the hardware components (e.g. frame grabber, robot controller card) are different compared to those used at the Johns Hopkins University. It is obvious, that the implementation of different hardware cards leads to the need for adapted control software. Therefore, the author developed two software libraries, the *robot class* and the *image class*, with implemented software drivers (low level control) and basic functionality for robot motion and image processing (compare section 6.6.2). Furthermore, all components of the client/server architecture, presented in section 7.3, are integrated on one PC-based workstation, the 'visual servoing workstation' which has already been presented in section 6.6. Both software libraries, *image class* and *robot class*, provide convenient communication between the visual servoing control and the hardware components.

Another issue which had been addressed in the new X-ray setup are image distortions. Distortions caused by the employed mobile C-arm system (Siemens Siremobil Iso-C) have been tested with a special calibration plate fixed on the image intensifier. As described in section 6.5.1, these tests demonstrated that the employed Siremobil C-arm system showed only very less distortions, even near the image border. Therefore, the grabbed X-ray images have directly been used for image processing without prior distortion correction.

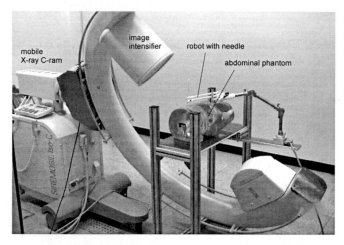

Figure 7.14: Experimental environment of the X-ray setup at the Siemens Basic Research Department. The new manipulator is rigidly attached to a radiolucent experiment table on which a human abdominal phantom is placed. X-ray imaging is provided by a mobile X-ray C-arm system (Siemens Siremobil Iso-C).

7.4.1 Preliminary Experiments with Metal Beads and Human Phantoms

The author installed the new X-ray setup in a laboratory at the Siemens Basic Research Department, Erlangen, Germany (see figure 7.14). A compact radiolucent experiment table was build up, which allows convenient imaging from various directions. For the first experiments several human phantoms (e.g. thorax, abdomen) have been used to verify the new test bed and the manipulator's characteristics. The employed phantoms have been hollow and allowed to place different targets inside. The new robotic manipulator with its passive arm is rigidly attached to the table, while the needle tip (rotation point) is placed on the phantom. X-ray imaging is provided by the mobile C-arm system Siemens Siremobil Iso-C. The system has an isocentric design, so that the C-arm can be moved around its two rotation axes, while its optical axis always goes through the rotation point [47]. The C-arm has an approximate source-to-intensifier distance of 1 meter, and an intensifier diameter of 25 centimeter. The goal of these preliminary trials was to prove the new robotic system in terms of accuracy, usability, and robustness.

Figure 7.15 shows a radiograph of the abdominal phantom together with the robot. The image gives an example, where precise needle segmentation is getting increasingly harder with the employed threshold approach. Detection of the needle or the guiding cannula via threshold requires that these structures have the lowest grayscale values in the image (compare section 6.5.2). But since the needle and the guiding cannula can be overlaid with other high contrast objects like bones or the spine, the metallic needle might not be automatically the darkest structure in the image. Therefore, the employed threshold approach for needle detection can get inappropriate in case of the presence of other high density structures in the image.

This problem could be reduced by defining a certain region of interest within which the needle is detected before starting the alignment procedure. It can be stated that under controlled conditions as in the presented experiments, this approach was still appropriate and lead to good results. However, for clinical use, more sophisticated image processing algorithms for reliable needle detection have to be implemented, which allow for e.g. edge detection or automatic segmentation [124][118].

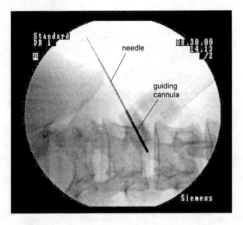

Figure 7.15:   X-ray radiograph of the robot's end-effector together with an abdominal phantom. Both, the guiding cannula and the needle are clearly visible in the image.

7.4.2 Experiments in Automatic Needle Alignment in the Optical Axis

Another automatic and visually controlled needle placement technique has been tested with the same X-ray setup: visual servoing with the *axial aiming technique*. As described in section 2.2 and 6.3.5 the purpose of this approach is to superimpose the needle and the target so that both show up in the image as a single point. This automatically leads to alignment of the needle with the target. After alignment has been achieved, depth control during needle insertion has to be performed in a side view by rotating the C-arm.

Purpose of the conducted experiments was to validate the stability of the alignment algorithm and to determine how many iterations are necessary for alignment within the X-ray beam. However, the accuracy in needle alignment has not been tested in the experiments, because this basically depends on how precise the robot (tip of guiding cannula) has been manually positioned above the target in the image. Therefore, all experiments have been conducted without using a target bead.

Figure 7.16 shows the stepwise needle alignment in several X-ray images. In order to achieve more realistic conditions a human abdominal phantom has been used for simulation of the puncture of the intervertebral disk between L3 and L4 (lumbar). The robot is positioned above the vertebra lumbales at the desired insertion point. The target and the insertion point are superimposed in the image. Then the robot automatically moves the needle stepwise into the optical axis under intermittent X-ray imaging. Figure 7.16-b shows the stepwise needle advancement during the alignment procedure.

(a)

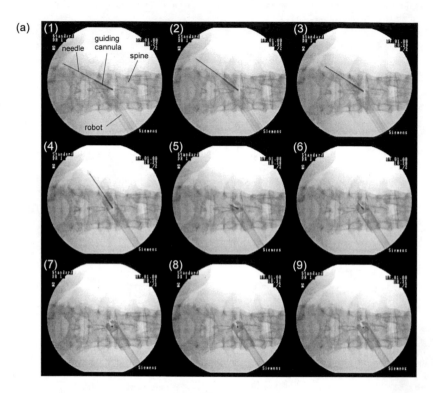

(b)

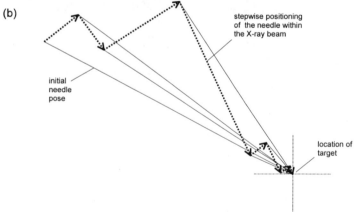

Figure 7.16: Simulation of the puncture of an intervertebral disk of a human abdominal phantom (disc between L3 and L4). The robot is positioned above the vertebra lumbales at the desired insertion point (a). Target area and the tip of the guiding cannula are superimposed in the image. Then the robot automatically rotates the needle stepwise into the X-ray beam under intermittent imaging. (b) Needle advancement during the alignment process.

7.4.3 Cadaver Study with Pig Organs

After successful implementation and testing of the visual servoing test bed with simple target beads and human phantoms, a number of cadaver trials have been performed. The goal was to prove the new manipulator and the image-guided needle placement approach under more realistic clinical conditions.

Fresh thoracic and abdominal pig organs have been organized from a local slaughterhouse. Figure 7.17 shows the pig cadaver consisting of two lungs, the heart, the liver and two separate kidneys. Legal regulations decree that all organs of a slaughtered animal have to be incised to prove its state of health and quality. Therefore, the test cadaver shows several deep cuts. To prevent the organs' surface from drying up and to facilitate handling of the cadaver during the experiments, it is put into a transparent plastic bag. For experiments the cadaver was used under room temperature for max. 2 hours continuously. Then it was cooled down again in a refrigerator for several hours. This let the cadaver spoil only slowly and allowed to use the same organs for several times.

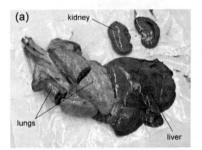 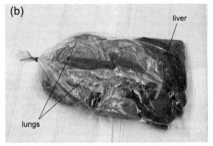

Figure 7.17: Cadaver study with thoracic and abdominal pig organs. (a) The organs' condition coming from a local slaughterhouse (organs of a slaughtered animal are always incised in order to verify its state of health and quality). (b) To prevent the organs' surface from drying up and to facilitate its handling during the experiments, the cadaver is put into a transparent plastic bag.

Figure 7.18 shows the cadaver setup during the automatic alignment procedure. The plastic bag with the pig organs is placed on the radiolucent experiment table. A metal bead with a diameter of 2 millimeters is implanted into one of the organs and served as target. After obtaining alignment with the implanted bead, the needle has been inserted manually.

In earlier experiments a *freestanding target bead* was used (compare figure 7.7 and 7.10), which allowed to directly observe the location of the needle tip on the way to the target. Furthermore, the absence of surrounding tissues avoids the deflection of the needle during insertion, which automatically leads to higher precision.

However, performing *needle insertion through real tissues* as in the cadaver study, the needle has to penetrate more or less tough material on the way to the target. Due to the asymmetric grinded needle tip, the needle may deflect during insertion especially in case of very tough tissues. This leads to inaccuracies in needle placement, although the needle was exactly aligned with the target before insertion. To minimize this problem, a special needle with a sharp and symmetric needle tip has been employed for the experiments.

A further problem results from the invisibility of the needle tip during insertion: the operator cannot directly see anymore how deep to insert the needle to reach the target bead. In case that the needle would be advanced too far, the target bead would get pushed aside while

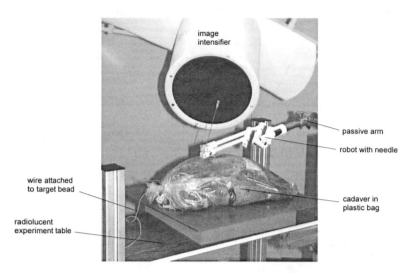

Figure 7.18: The covered test cadaver is placed on the radiolucent experiment table during the automatic needle placement procedure. A small metal bead, provided with a thin wire, is implanted in one organ and serves as target.

passing. This displacement of the bead would falsify the result. To get rid of this problem, the metal target beads were provided with a thin wire, which allows automatic electrical contact sensing to detect contact between the needle and the metallic target bead during needle insertion (compare figure 7.19). However, the experiments showed that inserting the needle stepwise - under intermittent fluoroscopic imaging - is closer to medical practice and ensures to avoid passing the target.

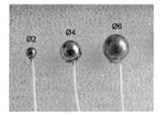

Figure 7.19: Target beads with different diameters for implantation into the cadaver. A thin wire is soldered to the bead and allows automatic electrical contact sensing to detect contact between the needle and the metallic target bead during needle insertion.

There have been several needle placement experiments conducted in the lungs, the liver, and the kidneys. Figure 7.20 shows two radiographs taken during these experiments. The target bead with its wire is clearly showing up in the images. After execution of the automatic alignment procedure, the needle is inserted stepwise while intermittent imaging provides the actual location of the needle tip. Supported by the sharp and symmetric tip, needle insertion can be performed under soft rotation of the needle between the operators fingers. The rotation of the needle during insertion facilitates penetration of tough tissues and avoids needle deflection.

The experiments showed, that depending on the type and density of the tissue, the so-called *windowing* of the digital radiograph (adjustment of brightness and contrast) is important to make certain structures visible. The radiograph shown in figure 7.21 demonstrates the differences in tissue density resulting from the grade of X-ray absorption. The heart and the liver are comparatively dense soft tissues, while the lungs show up much lighter in the X-ray image due to the amount of air in the tissue.

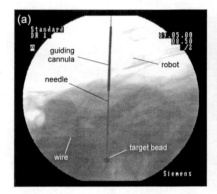 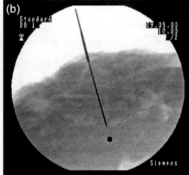

Figure 7.20: Two radiographs taken during the cadaver trials in X-ray guided needle placement: lung puncture (a) and puncture of the liver (b). The target bead with its wire is clearly showing up in the images. After execution of the automatic alignment procedure, which detects basically the guiding cannula in the image, the needle is inserted stepwise by hand under intermittent imaging.

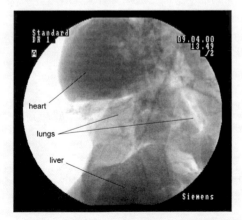

Figure 7.21: This radiograph demonstrates the differences in tissue density. The heart and the liver are comparatively dense soft tissues, while the lungs show up much lighter in the X-ray image due to the amount of air in the tissue.

7.5 Results in Automatic X-ray Guided Needle Placement

7.5.1 Experiments and Results in Depth Estimation

The mathematical background of depth estimation after X-ray guided needle alignment has been presented in section 6.3.1, Step IV. It is based on the computation of the cross-ratio of at least 4 corresponding points in the image and in Euclidean space, which have to be aligned on a straight line (see figure 7.22-a). In perspective projection cross-ratios are invariant and can therefore be used to determine the desired 3D distance $\|ET\|$ between the needle tip and the target.

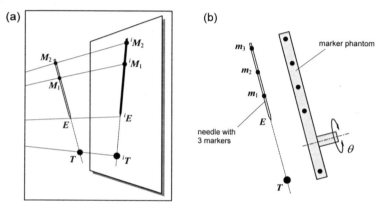

Figure 7.22:   According to projective geometry, the cross ratio $[^iM_1, {}^iM_2, {}^iE, {}^iT]$ of the image points is equal the cross ratio of the corresponding 3D points $[M_1, M_2, E, T]$. (a) After 3D alignment of the needle, the required insertion depth can be estimated using cross-ratios between characteristic points on the needle. (b) For initial experimental verification of the insertion depth a marker phantom with 5 beads has been employed.

However, following conditions have to be given for the computation of $\|ET\|$: (i) the Euclidean distances between the marker beads and the skin entry point on the needle have to be known; and (ii) the digital radiograph must provide at least two marker beads M_i, the needle tip E (entry point) and the target T. In order to increase precision in computation of $\|ET\|$ more marker beads than M_1 and M_2 can be used.

In preliminary experiments the 'achievable' accuracy of the depth estimation approach has been verified with a special marker phantom (see figure 7.22-b). The marker phantom consists of a small acrylic bar provided with 5 metallic marker beads (Ø4mm) precisely positioned in a straight line with defined distances. As shown in figure 7.22-b, the marker phantom corresponds to a needle provided with 3 additional marker beads.

Obviously, in medical practice it is not appropriate to use a puncture needle provided with any marker beads. Therefore, either the guiding cannula can be provided with at least one bead[24] for computation of the cross-ratio, or a needle phantom with marker beads can be used

---

[24] The guiding cannula already provides two intrinsic points, the tip and the end point of the cannula. Adding one marker bead is sufficient to compute the cross-ratio (compare figure 7.22-a).

for the automatic alignment and depth estimation procedure. Afterwards, the needle phantom might be replaced by a real needle for puncture.

The five beads of the marker phantom are positioned very accurately at predefined locations. This allows to determine the precision of this depth estimation approach very accurately under idealized conditions. Furthermore, the marker phantom can be tilted around a perpendicular axis to determine the precision in computation of $\|ET\|$ depending on the tilt angle θ.

Using the presented marker phantom with 3 additional markers m_i, the insertion depth $\|ET\|$ is computed in a least-square sense as follows: first, the cross ratios λ_n can be computed for all possible combinations of marker points. In case of the needle phantom with its five marker beads, four cross ratios can be derived:

$$\begin{aligned}
\lambda_1 &= [{}^iT, {}^iE, {}^im_1, {}^im_2] \\
\lambda_2 &= [{}^iT, {}^iE, {}^im_1, {}^im_3] \\
\lambda_3 &= [{}^iT, {}^iE, {}^im_2, {}^im_3] \\
\lambda_4 &= [{}^iT, {}^im_1, {}^im_2, {}^im_3]
\end{aligned} \tag{7.1}$$

Equation (6.3), which determines $\|ET\|$, can be formulated as

$$\|ET\|_n = \frac{a_n}{b_n} \quad \text{with} \quad \begin{aligned} a_n &= \left[\lambda_n \cdot \|E\,M_1\| \cdot \|E\,M_2\|\right] \\ b_n &= \left[\|M_2\,M_1\| - \lambda_n \cdot \|E\,M_1\|\right] \end{aligned}, \quad n = 1..4 \tag{7.2}$$

which can be solved in a least-square approach [9] as follows

$$\|ET\| = \left(b^T b\right)^{-1} b^T a \quad \text{with} \quad \begin{aligned} a &= \left(a_1 \; a_2 \; ... \; a_n\right)^T \\ b &= \left(b_1 \; b_2 \; ... \; b_n\right)^T \end{aligned}. \tag{7.3}$$

Digital radiographs have been taken of the needle phantom with a prototypical C-arm system equipped with a solid state detector. This detector provides digital X-ray images *without distortions*. Figure 7.23-b shows several radiographs of the needle phantom with different tilt angles θ. These experiments in depth estimation demonstrate the precision of this approach depending on θ. The ensuing computation of the insertion depth $\|ET\|$ results in different deviations δ. Table 7.1 presents the results of $\|ET\|$ and δ with this least-square approach.

As expected, the depth estimation is getting inaccurate the more the needle phantom is tilted. This is because the inaccuracies in determination of the marker positions have more influence on the result the smaller the measured marker distances in the images are. Further-

θ	$\|{}^iT{}^iE\|$	$\|{}^iE{}^iM_1\|$	$\|{}^iM_1{}^iM_2\|$	$\|{}^iM_2{}^iM_3\|$	$\|TE\|_{\lambda_1}$	$\|TE\|_{\lambda_2}$	$\|TE\|_{\lambda_3}$	$\|TM_1\|_{\lambda_4}$	$\|TE\|$	δ
0°	325.14	130.18	129.87	130.33	49.75	49.92	50.19	70.68	49.99	-0.01
20°	303.26	123.70	126.00	126.92	50.63	50.30	49.78	68.87	50.14	0.14
40°	247.84	103.25	105.38	108.76	49.71	50.03	50.53	71.45	50.18	0.18
60°	159.76	68.07	69.29	72.38	48.37	49.08	50.23	72.39	49.41	-0.59
80°	53.26	23.22	22.80	24.37	44.55	46.68	50.31	67.94	47.68	-2.32

Table 7.1: Estimation of the distance $\|TE\|$ between the 'target bead' and the 'insertion point bead' for different inclination angles θ of the marker phantom with a least-square approach. Notice: $\|TE\|_{\lambda_4} = \|TM_1\|_{\lambda_4} - 20 \, \text{mm}$

more, these results demonstrate that already one additional marker bead improves the accuracy of the depth estimation significantly.

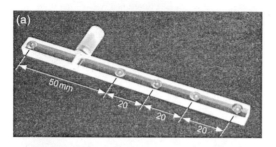

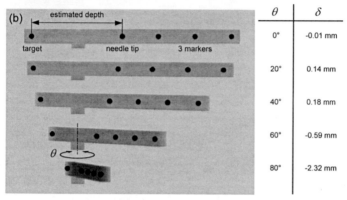

θ	δ
0°	-0.01 mm
20°	0.14 mm
40°	0.18 mm
60°	-0.59 mm
80°	-2.32 mm

Figure 7.23: Experiments in depth estimation using cross ratios. (a) The acrylic marker phantom provides five metal marker beads (Ø4mm) precisely positioned in a line with defined distances. The marker beads symbolize the target, the insertion point, and 3 additional markers. (b) X-ray images showing the test phantom in five different inclinations (angle θ). The ensuing computation of the distance between the 'target bead' and the 'insertion point bead' results in different deviations δ, according to table 7.1.

7.5.2 Accuracy Verification After Needle Alignment

The user interface presented in section 6.3.4 has integrated a special functionality for verification of the remaining deviation after needle alignment. To determine the 3D deviation vector Δ between the needle axis and the target midpoint (see figure 7.24), *two verification images* of the needle from two known viewpoints have been taken after the alignment process was accomplished. In these two images the distances Δ_1 and Δ_2 between needle axis and target midpoint was determined. Additionally, the diameter of the target bead in both images has been measured in terms of pixels. Since the real diameter of the bead is known in millimeters, the relation η between millimeter and pixel size could be easily computed. It is assumed that η_1 and η_2 of both images are constant near the target bead. With this assumption it is straightforward to compute the 3D deviation vector Δ in the needle coordinate frame n.

This functionality is implemented in the user interface presented in section 6.3.4. The algorithm requires both viewing directions (optical axes) lying in a horizontal plane. Therefore, in case of *CCD-imaging*, the camera is just moved on the experiment table (horizontally) to provide both views (compare figure 7.5). Markers on the table support in orienting the camera in predefined viewing angles. For *X-ray imaging*, the C-arm is positioned horizontally for acquisition of both images in a horizontal plane (compare figure 7.11). In this case, the viewing angle can be directly determined by a scale on the C-arm. In case of multi target alignment, the automatic computation of the remaining deviation vector can be computed for each target bead.

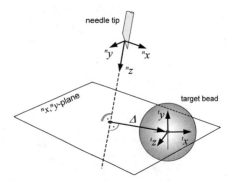

Figure 7.24: Remaining 3D deviation vector Δ between the needle axis and the target midpoint after the alignment process. The deviation vector is computed with regard to the needle coordinate frame $[^n x, ^n y, ^n z]$.

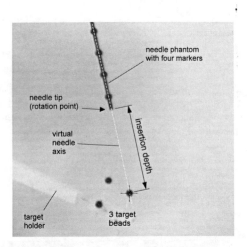

Figure 7.25: CCD-image taken during the automatic needle alignment procedure with the C-arm-simulator. A black needle phantom provided with four marker beads is employed for depth estimation.

7.5.3 Results with CCD-imaging

During experiments using CCD imaging, several black target beads with a diameter of 4 millimeters have been used (compare figure 7.25). The distance between needle tip and target varied between 52 and 61 millimeters. A total of 60 automatic CCD-guided alignments have been conducted. After needle alignment, the remaining 3D deviation vector Δ between the needle axis and the target midpoint is determined by two verification CCD-images taken from known viewpoints (compare section above). Both images have to show the needle together with the target bead in order to compute the 3D deviation vector Δ.

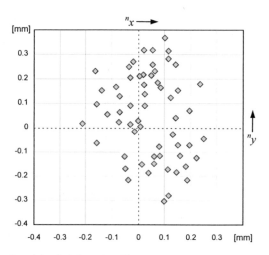

Figure 7.26: Remaining deviation Δ in millimeters between needle axis and target midpoint after alignment (60 trials). The diagram illustrates the $^{n}x,^{n}y$-plane of the needle coordinate system (compare figure 7.24). The points in the diagram represent the location of the target midpoint after each alignment trial.

Figure 7.26 shows a dot diagram of the remaining deviation vector Δ after automatic needle alignment. The diagram illustrates the $^{n}x,^{n}y$-plane of the needle coordinate system (compare figure 7.24). The points in the diagram represent the location of the target midpoint after each alignment trial. The procedure typically needed 3 to 5 iterations to achieve alignment in one plane π.

The statistical evaluation is presented in table 7.2. The mean and maximum deviation were 0.20 and 0.38 millimeters. Considering the stopping criterion of the visual feedback loop, $\delta = \gamma_d - \gamma < 0.2°$ for rotation of the needle in plane π (compare section 6.3.3), the mean deviation of 0.20 mm is approximately equivalent to the deviation caused by this angle of 0.2° and an insertion depth of 60 millimeters: $\tan 0.2° \cdot 60$ mm = 0.2094 mm.

$|\Delta| = \sqrt{^{n}x^2 + {}^{n}y^2}$	target location	no. of trials i	mean of $|\Delta|_i$ [mm]	max. of $|\Delta|_i$ [mm]	stand.dev. of $|\Delta|_i$ [mm]	depth [mm]
◇ CCD-imaging (JHU)	freestanding bead	60	0.20	0.38	0.08	52..61

Table 7.2: Remaining deviation $|\Delta|$ between the needle axis and the target bead after automatic needle alignment (60 trials, compare figure 7.26).

In these trials the deviation is determined *after needle alignment*, not after needle insertion (see figure 7.25). Therefore, a possible impact of e.g. needle bending or a non-coaxial needle translation during insertion is eliminated. These idealized conditions may be the reason for the very accurate results in the CCD-imaging experiments. However, it demonstrates that the presented approach in automatic needle alignment itself is very accurate without further interference by e.g. a non-coaxial needle, needle bending due to the insertion process, or target motion during the procedure.

After each needle alignment trial the system automatically determines the required insertion depth as described in section 6.3.1, Step IV. For this purpose the employed needle phantom has been provided with four marker beads with known distance in between (see figure 7.25). These trials have been performed with three different target beads for twenty times. The error in depth estimation is shown in figure 7.27. The real depth of each target bead is indicated by the dotted lines.

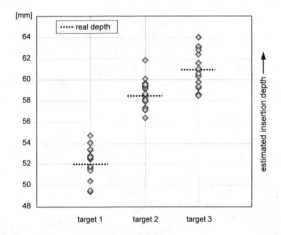

Figure 7.27: Determined insertion depth after needle alignment for three different target beads. The dotted lines indicated the real depth of each target.

Table 7.3 presents the statistical evaluation of the deviations in figure 7.27. Although the points in the figure vary a lot (about ±2mm), the mean error in the estimated insertion depth is only 0.29 mm. The large standard deviation emphasizes the relatively large variation of the estimated depths.

	real depth [mm]	estimated depth (mean) [mm]	mean error [mm]	error range [mm]	standard deviation [mm]
target 1	52.0	52.23	0.23	-2.58 .. 2.75	1.37
target 2	58.5	58.85	0.35	-2.13 .. 3.35	1.22
target 3	61.0	60.72	0.28	-3.55 .. 2.00	1.64
			mean: 0.29		

Table 7.3: Error in depth estimation after automatic needle alignment. Although the mean error in each target is quite small, the large standard deviation shows that the determined insertion depth varies a lot (about ±2mm, compare figure 7.27).

7.5.4 Results with X-ray-imaging

Initial X-ray guided needle placement experiments have been performed at the Johns Hopkins University (JHU) with the RCM robot and a small X-ray C-arm (compare figure 7.11). In these experiments a metal bead with a diameter of 2 millimeters has been employed as target. Two cases were investigated: (i) needle placement with a freestanding target bead without surrounding tissues, and (ii) needle placement with the bead implanted in a single pig kidney. For both cases the needle insertion depth was about 70 millimeters.

Comparable experiments were conducted at the Siemens Basic Research Department with the novel needle guiding robot. In initial trials a freestanding metal bead with a diameter of 4 millimeters served as target. Further experiments in automatic needle placement were performed with the same target bead implanted in different organs of a pig cadaver. Here, the needle insertion depth has been at least 80 millimeters.

All results achieved in X-ray guided needle placement experiments at the Johns Hopkins University and the Siemens laboratories are presented and compared in the following two sections.

Results with a freestanding target bead

After automatic alignment the needle is advanced close to the target bead. Figure 7.28 shows some verification radiographs taken from two known viewpoints (first and second view) after placement of the needle. In both verification images the deviations Δ_1 and Δ_2 between needle axis and target midpoint are determined and used for computation of the 3D deviation vec-

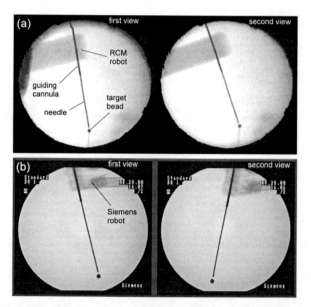

Figure 7.28: X-ray guided needle placement with a freestanding target bead without surrounding tissues. (a) Verification radiographs from test series performed at the Johns Hopkins University; bead diameter 2mm, insertion depth about 70mm. (b) Verification radiographs from test series performed at the Siemens laboratories; bead diameter 4mm, insertion depth about 85mm.

tor Δ. The remaining deviation Δ after 13 automatic X-ray guided needle placement trials is shown in figure 7.29. The diagram illustrates the $^{n}x,^{n}y$-plane of the needle coordinate system while the points represent the location of the target midpoint after each needle placement trial.

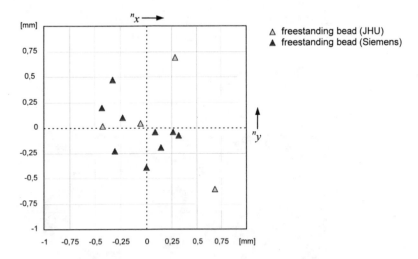

Figure 7.29:   Remaining deviation Δ after X-ray guided needle placement with a freestanding target bead. Experiments have been performed at the Johns Hopkins University ($\triangle$) and at the Siemens Medical Solutions R&D laboratories ($\blacktriangle$). The diagram illustrates the $^{n}x,^{n}y$-plane of the needle coordinate system (compare figure 7.24). The points in the diagram represent the location of the target midpoint after each needle placement trial.

Table 7.4 shows a statistical evaluation of the remaining deviation $|\Delta|$ between the needle axis and the target bead after needle placement. While the accuracy in the test series performed at the Johns Hopkins University is about ±1mm, in the similar experiments performed at the Siemens laboratories an accuracy of approximately ±0.6mm. This is mainly due to the image quality of the used X-ray C-arm (compare figure 7.28). While at the Johns Hopkins Medical School a quite old X-ray C-arm with a small image intensifier has been used, a new sophisticated C-arm system was employed for the test series at the Siemens laboratories.

$|\Delta| = \sqrt{^{n}x^{2} + ^{n}y^{2}}$		target location	no. of trials i	mean of $|\Delta|_{i}$ [mm]	max. of $|\Delta|_{i}$ [mm]	stand.dev. of $|\Delta|_{i}$ [mm]	depth [mm]
$\triangle$	X-ray imaging (JHU)	freestanding bead	4	0.54	0.92	0.37	ca. 70
$\blacktriangle$	X-ray imaging (Siemens)	freestanding bead	9	0.33	0.57	0.14	ca. 85
		all trials:	13	0.39			

Table 7.4:   Results in X-ray guided needle placement with a freestanding target bead. Statistical evaluation of the remaining deviation $|\Delta|$ between the needle axis and the target bead after needle placement.

Results with implanted beads in a pig cadaver

A more clinical scenario has been achieved in further test series using different pig organs for needle puncture. Again, a small metal bead served as target, which has been placed at different locations within the cadaver. This bead showed up very clearly as defined, round spot in the X-ray radiographs. Although, this does not reflect the structure of a realistic anatomical target, it is the easiest way to determine precisely the principle accuracy of the needle placement approach within body tissues. However, at the end of this section, the puncture of a contrasted pig kidney is presented and problems in verifying the accuracy in this experiment is discussed.

Needle placement has been performed within different organs of the pig cadaver: in the lungs, the liver, the heart, and in the kidneys (compare section 7.4.3). The kidneys have been used separately and not embedded within other tissues.

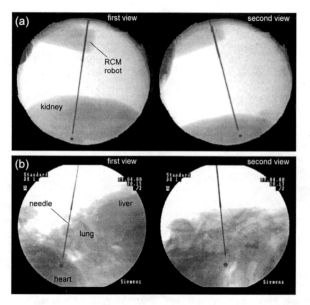

Figure 7.30:   X-ray guided needle puncture within different pig organs. A metal bead has been implanted and served as puncture target. (a) Verification radiographs from 'kidney' test series performed at the Johns Hopkins University; bead diameter 2mm, insertion depth about 70mm. (b) Verification radiographs from 'lung' test series performed at the Siemens laboratories; bead diameter 4mm, insertion depth about 80mm.

Figure 7.30 shows verification radiographs taken after automatic needle placement in different test series. Again, the radiographs show significantly different image quality due to the different C-arm systems. The method of determining the remaining 3D deviation vector Δ is equivalent to that described above.

The diagram in figure 7.31 summarizes the results in the needle placement trials within different pig organs. Again, the $^nx,^ny$-plane of the needle coordinate system is shown in the diagram while its dots represent the location of the target midpoint after each needle placement trial. At the Johns Hopkins Medical School only kidneys have been punctured.

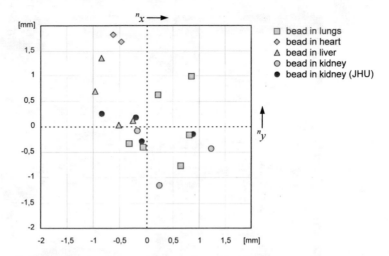

Figure 7.31: Remaining deviation Δ after X-ray guided needle placement within different pig organs. Experiments have been performed at the Johns Hopkins University ($\bullet$) and at the Siemens Medical Solutions R&D laboratories ($\square\diamond\triangle\circ$). The diagram illustrates the ^nx,^ny-plane of the needle coordinate system (compare figure 7.24). The points in the diagram represent the location of the target midpoint after each needle placement trial.

The overall accuracy in these cadaver trials is about ±1.5mm. The puncture trials in the heart show the largest deviation of about 2mm. The heart muscle is the toughest tissue of the employed cadaver. This may explain the relatively large deviation due to needle bending and drift during passing these tough tissues. The relatively small deviation in the kidney puncture performed at the Johns Hopkins Medical School may have its reason in the smaller insertion depth and shorter path through the kidney tissue (see figure 7.30-a).

$\|\Delta\| = \sqrt{^nx^2 + ^ny^2}$		target location	no. of trials i	mean of $\|\Delta\|_i$ [mm]	max. of $\|\Delta\|_i$ [mm]	stand.dev. of $\|\Delta\|_i$ [mm]	depth [mm]
$\square$	X-ray imaging (Siemens)	bead in lungs	6	0.85	1.52	0.39	70..90
$\diamond$		bead in heart	2	2.02	2.26	0.34	
$\triangle$		bead in liver	4	1.08	1.87	0.74	
$\circ$		bead in kidney	3	1.00	1.56	0.68	
$\bullet$	X-ray imaging (JHU)	bead in kidney	4	0.59	0.91	0.14	ca. 60
		all trials:	15	1.25			

Table 7.5: Results in X-ray guided needle puncture within different pig organs. Statistical evaluation of the remaining deviation $|\Delta|$ between the needle axis and the target bead after needle placement.

Finally, a needle placement test series is presented with a contrasted pig kidney. This is leading to very realistic puncture conditions with a more or less indistinct target area. This signifi-

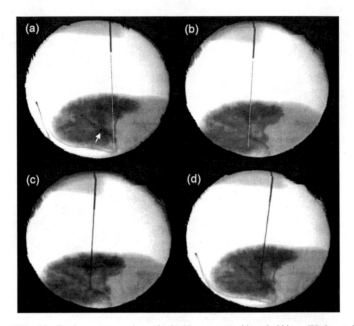

Figure 7.32: Needle placement experiment in the X-ray set-up with a pig kidney. We inserted contrast medium into one arteria renalis interlobares of the pig kidney, and decided to
puncture the first main ramification (see white arrow) since this point was clearly visible on the radiograph. The frayed border is caused by the distortion correction algorithm. (a) Starting position of the needle-guide. The virtual needle axis is drawn onto
the image. (b) Final orientation of the needle-guide after complete alignment procedure.
(c) Needle is successfully inserted into the target structure through the needle-guide.
Here insertion depth was controlled by the physician on the screen. (d) Verification of
puncture with first viewing direction.

cantly complicates the determination of the most appropriate needle trajectory to the target as
well as the verification of the remaining deviation after puncture.

Contrast medium has been injected into one arteria renalis interlobares. As puncture target
served the first main ramification of the contrasted artery since this point was quite clearly
visible on the radiograph. Figure 7.32-a shows the starting orientation of the guiding cannula
before the automatic alignment procedure started. The catheter for insertion of the contrast
medium is visible in the lower left corner. The virtual needle axis is drawn onto the image
during the procedure so that the physician can easily verify the procedure. The final needle
orientation after the complete alignment procedure is given in image 7.32-b. Then, the needle
has been inserted manually through the needle-guide, while the insertion depth was controlled
on the screen under intermitted X-ray imaging. Finally, two verification images from known
viewpoints have been taken for verification of the remaining deviation after puncture (see
figure 7.32-c, d). However, in this case the definition of a 3D deviation vector Δ is difficult
because neither the target point nor the scale close to the target in both verification images can
be precisely determined. Even the definition of the corresponding target points in both views
is difficult and source of inaccuracies in the puncture procedure, especially in case of a not
clearly visible target.

While the use of a round target bead with a known diameter easily allows to estimate the scale by computation of the ratio between the real and the measured diameter in the image (mm/pixel), this is not possible in the puncture of normal anatomical targets. Nevertheless, even in case of the puncture of a contrasted pig kidney the scale has been tried to determine by comparison of the deviation with the needle diameter in both verification images. The achieved accuracy has been estimated at about ±2mm.

Chapter 8
Needle Placement Experiments Using CT-imaging

This chapter describes the setup and experiments in automatic image-guided needle placement using CT-imaging. The theoretical framework of this approach has been described in section 6.4. All experiments were performed with a Somatom Plus4 single slice CT scanner (Siemens AG, Germany) at the Siemens CT laboratories in Forchheim, Germany. To achieve fast image acquisition the 'CARE Vision' mode has been used (i.e. CT-fluoroscopy, compare paragraph 2.3.2), which allowed an image sample rate of up to 8 images per second. These reconstructed images are provided by the system via video output as NTSC signal. All experiments presented in this chapter have been performed with the smallest CT slice thickness of 1 millimeter, in order to achieve maximum precision in needle placement.

The following paragraph gives a comprehensive overview of the robotic system setup together with the CT-scanner. In order to determine the achievable accuracy in robot-guided needle placement with the novel image controlled approach, several puncture test series have been performed on fruits and different pig organs. The results are presented in the ensuing paragraphs.

8.1 System Design for CT-guided Needle Placement

The two main components of a CT unit are the actual examination unit, the so-called *gantry*, and the *patient table* (compare figure 8.1). Both are installed in the CT examination room. The system is controlled and operated by the *CT console*, which is placed outside the examination room [79].

Before scanning, the patient is placed on the table top of the patient couch. The operator moves the patient into the scan plane inside the CT gantry by pressing buttons on the user panel on the gantry. Afterwards, the operator is leaving the examination room before scanning is started in order to avoid radiation exposure. Adjustment of the scan parameters and scan release is performed from outside the examination room at the CT console.

In case that the radiologist wants to perform image-guided interventions within the CT gantry, for example in case of CT-guided biopsies, common CT systems provide a *CT-fluoroscopy scan mode* (compare section 2.3.2). This scan mode allows a higher image sample rate[25], displayed on an in-room monitor. Furthermore, basic operation of the CT system can be performed from inside the examination room (see figure 8.1): the operator can turn scanning on and off via foot pedal. Furthermore, a joystick attached to the patient table provides convenient horizontal repositioning of the patient within the scan plane during the intervention. Additionally, the CT images in fluoroscopy mode are provided by the CT console as NTSC video signal for documentation and recording with a video recorder.

---

[25] The higher image sample rate of up to 8 images per second is achieved by reducing the image resolution. While in common CT scanning an image matrix of 512x512 is reconstructed, CT fluoroscopy computes only an image matrix of 256x256 [79]. However, this reduced image resolution is sufficient for guiding in interventional procedures.

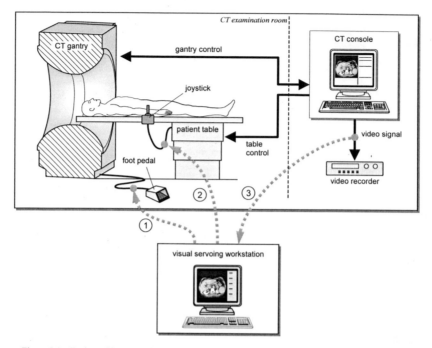

Figure 8.1: Basic architecture of the experimental CT setup for automatic needle placement. The entire CT control and data acquisition for image reconstruction is performed by the CT console computer. In the CT-fluoroscopy scan mode the operator can control scanning via foot pedal. Horizontal table motion is achieved by using the table joystick. The visual servoing workstation adopts basic control functions by simulating the foot pedal (1) and the joystick (2). Image acquisition is realized via frame grabber, which is connected to the video output at the CT console (3).

The visual servoing workstation has been slightly modified for the CT-guided needle placement experiments. Both software libraries for image acquisition and robot control (*image class*, *robot class*) have been supplemented and adapted for this new imaging modality (compare section 6.4.2 and 6.6.2). A schematic of the basic architecture of the CT test bed is shown in figure 8.1. It demonstrates the interfaces and communication between the CT system and the visual servoing workstation.

The entire CT control and data acquisition for image reconstruction is performed by the CT console (host computer). A video output provides CT fluoroscopy images for the visual servoing workstation. The video signal is directly captured via frame grabber and is digitally processed by the *image class*.

Just as in the X-ray setup discussed before, the visual servoing workstation again employs the foot pedal interface for remotely controlled scan release. Pressing the foot pedal closes an electric circuit and indicates the CT scanner to start scanning. Using this interface, the visual servoing workstation simulates the signal by a relay integrated on the axis encoder board (compare 6.6.1).

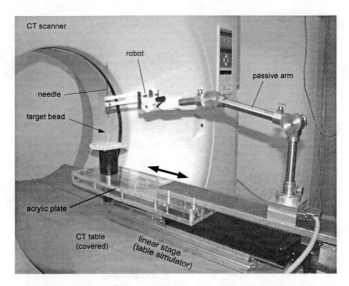

Figure 8.2: CT robot-setup with table simulator and target bead. The robot is fixed to a passive arm, which is mounted on a linear drive (table simulator) to perform translational motions of the robot and the target. In preliminary experiments a freestanding metal bead is employed as target structure.

However, a major problem in this setup was to realize a control of the *horizontal CT table motion*, which would require a communication of the visual servoing workstation with the internal CT CAN-bus system. But to realize this within an acceptable period of time was not possible for the author. In preliminary tests this problem has been solved by simulation of CT joystick functions. However, this approach did not provide appropriate control performance, and the author decided to build up a special '*CT-table simulator*', which has already been presented section 5.6. The simulator consists of a linear stage on which the robot and an acrylic plate are fixed (see figure 8.2). The acrylic plate is radiolucent and served as seat for the test objects. The linear stage allowed to perform translational motions of the test object and the robot within the scan field with excellent control performance and precision. The table simulator is placed on the CT table top as shown in figure 8.2.

Figure 8.3 shows the CT test bed for verification of the new automatic needle placement approach at the Siemens CT laboratories in Forchheim, Germany. The robot setup is installed at a Somatom Plus4 scanner in a CT test cabin. The visual servoing workstation is located close to the CT console outside the cabine.

8.2 Experiments and Results with CT-imaging

Evaluation of the novel approach in automatic CT-guided needle placement has been performed at the CT laboratories of the Siemens Computed Tomography division, Forchheim, Germany. During the experiments CT imaging has been provided by a Somatom Plus4 single slice CT scanner (Siemens AG, Germany).

The scanner was provided with a standard CT-fluoroscopy mode (CARE Vision CT™), which allowed an increased image sample rate of up to 8 images per second at a rotation

Figure 8.3:   Test bed for verification of automatic CT-guided needle placement at the Siemens CT
laboratories in Forchheim, Germany. Inside the X-ray cabin two CT scanners are installed, while the back one is used for the needle placement experiments. Outside the
cabin both the CT console and the visual servoing workstation are located.

speed of one second. The higher sample rate is achieved by reducing the image resolution.
While in common CT scanning an image matrix of 512x512 is reconstructed, CT fluoroscopy
computes only an image matrix of 256x256 [79]. However, this reduced image resolution is
sufficient for guiding in interventional procedures. The tube parameters during scanning were
adjusted to 50mA at 120 kV. Fluoroscopic images were acquired with a slice thickness of one
millimeter.

Initial experiments with a freestanding metal target bead have been performed to verify the
principle accuracy of this new CT-guided approach. Following experiments were conducted
with several fruits and finally with a pig cadaver.

8.2.1 Preliminary Experiments with a Freestanding Target Bead

A first assessment of the accuracy in automatic CT-guided needle placement has been made
with a small freestanding metal target bead (Ø2mm) without any surrounding tissue between
the guiding cannula and the bead. Figure 8.4 shows the setup within the CT scanner. The purpose of these tests was to evaluate the principle accuracy of both approaches in automatic
needle alignment described in section 6.4, without the influence of needle bending or needle
drift caused by surrounding tissues. The target bead is positioned on the acrylic plate of the
table simulator, which is placed on the CT table top. The distance between robot and target
bead was approximately 7 centimeters. After the alignment process has been accomplished
the needle is 'inserted' remotely controlled (see figure 6.19-11) and placed close to the target.

If the needle is *positioned within the CT scan plane* the remaining 3D deviation vector is
determined with the help of a CCD-camera (see figure 8.5). Two digital CCD-images of the
needle and the target bead have been taken from two known viewing directions. An angle
gage is fixed in a well defined orientation close to the target bead and shows up in both CCD
images (see figure 8.5-b,c). Thus, the particular horizontal viewing angle can be easily determined in the coordinate frame of the angle gage. After extraction of the needle axis and the

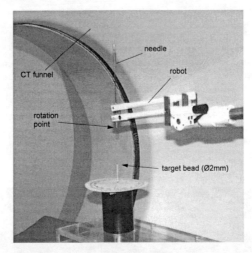

Figure 8.4: Automatic CT-guided needle placement experiments using a single freestanding metal
bead as target. This setup helped to determine the principle accuracy in needle placement which can be achieved with this approach (compare figure 8.2). The distance
between the target and the tip of the guiding cannula was about 7 cm.

target location in both camera images the remaining 3D deviation vector between needle axis
and target center can be easily computed. Both the needle and the target are painted black for
easy segmentation in the CCD-images. After analyzing these two camera images the desired
deviation vector Δ could be precisely computed.

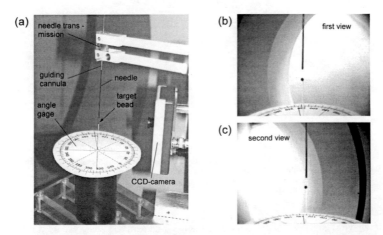

Figure 8.5: Accuracy verification after automatic CT-guided needle alignment. (a) Digital images
of the 'inserted' needle and the target bead are taken with a CCD-camera. After analysis of two CCD-images taken from two known viewing directions (angle gage), the
remaining 3D deviation vector can be precisely computed (b)(c).

In most alignment trials the target bead and the needle insertion point have not been located in the same CT scan plane. Therefore, needle placement *oblique to the scan plane* has been performed. However, this does not allow to observe the target and needle in one scan plane during the insertion process. Figure 8.6 shows four CT scans of different states of needle advancement. As described in section 6.4, scenario (b), the visual servoing workstation automatically moves the CT-table (or table simulator) in order to keep the needle tip permanently in the CT scan plane during remotely controlled insertion. This ensures that the tip is visible in the CT image all the time, even for oblique insertions. The cross section of the needle tip shows up as bright oval dot in the image. Furthermore, the desired insertion trajectory and the predetermined needle pose are projected onto the image. Thus, the physician can observe the deviation on the monitor between desired and actual needle tip position during remotely controlled insertion (compare figure 6.15-c).

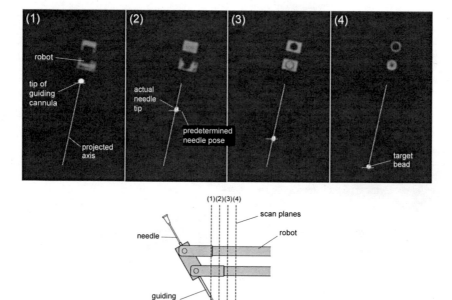

Figure 8.6:   CT-guided needle placement tilted to the scan plane. The CT images show different scans through the needle. The small horizontal line close to the round needle mark in the image shows the predetermined needle location. In the ideal case this line should be precisely subjacent the needle mark.

Here, accuracy verification is not performed with a CCD-camera. CT scans are used for computation of the 3D deviation vector Δ between the target bead and the needle axis. The principle is demonstrated in figure 8.7. After automatic needle placement, 4 CT scans were acquired in different table positions. The first scan contains the target bead T, while the last one shows the needle entry point E. The second and third scan represent cross-sections of the inserted

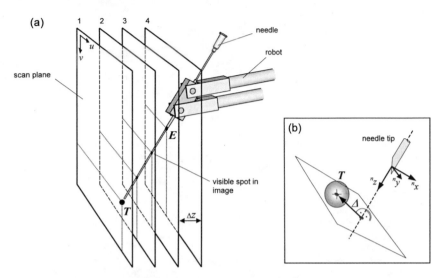

Figure 8.7: Accuracy verification in case of tilted needle insertion. (a) After needle placement, 4 control scans were acquired with different table positions. The first scan contains the target bead T, while the last one shows the needle entry point E. The second and third scan represent cross-sections of the inserted needle. All these cross-sections show up in the image as a bright spot. Now, the needle axis can be easily determined by a line fit computed for the needle cross-sections in the different scan planes. (b) Finally, the 3D deviation vector between the target bead and the determined needle axis can be precisely computed.

needle. All these cross-sections show up in the image as bright spots (compare figure 8.6). Since the table translations Δz between the scan planes are known, the needle axis can be easily determined by a line fit computed for the needle cross-sections in the different scan planes (compare figure 8.7-a). Finally, the 3D deviation vector between the target bead and the determined needle axis can be precisely computed (figure 8.7-b).

Preliminary experiments have been performed with a freestanding target bead with a diameter of 2 millimeters. The two scenarios were investigated: (i) needle placement within the CT scan plane, and (ii) needle placement tilted to the CT scan plane. The results of both test series are shown in table 8.6 and figure 8.8. In case of tilted needle placement, the tilt angle was about 65° to the vertical.

$|\Delta| = \sqrt{{}^n x^2 + {}^n y^2}$		target location	no. of trials i	mean of $|\Delta|_i$ [mm]	max. of $|\Delta|_i$ [mm]	stand.dev. of $|\Delta|_i$ [mm]	depth [mm]
$\triangle$	insertion in scan plane	freestanding without tissues	3	0.26	0.36	0.13	ca. 80
$\diamondsuit$	insertion tilted to scan plane	freestanding without tissues	19	0.59	1.10	0.28	78..88
		all trials:	22	0.54			

Table 8.6: CT-guided needle placement with a freestanding metal target bead with a diameter of 2mm. In case of tilted needle placement the mean tilt angle was 65° to the vertical.

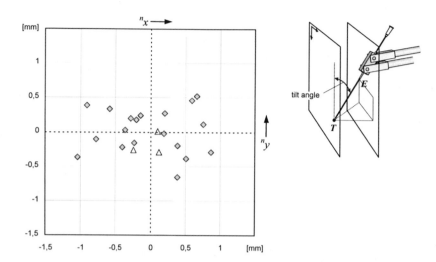

Figure 8.8: CT-guided needle placement with a freestanding target bead (Ø2mm). The diagram shows the remaining deviation Δ between needle axis and target midpoint after automatic needle alignment (see figure 8.7-b). Test series have been performed in needle placement within the scan plane ($\triangle$) and tilted to the scan plane ($\diamond$). In the latter case the tilt angle was about 65° to the vertical.

In case of needle placement within the CT scan plane the insertion depth was about 80 millimeters. Three trials have been conducted. The remaining deviation was determined by two CCD-images for verification (compare section above). With a mean deviation of only 0.26 millimeters the alignment process is very accurate. Nineteen needle placement trials have been performed tilted to the CT scan plane. Here, the insertion depth varied between 78 and 88 millimeters. After needle alignment the remaining 3D deviation vector Δ between the needle axis and the target midpoint was determined by several verification scans as described above (compare figure 8.7). In these trials the mean deviation after needle placement was 0.59 millimeters.

It is not very surprising, that the remaining deviation is much smaller for needle placement within the CT scan plane than tilted to it. This is basically because in scenario 1 the target bead and the robot's needle tip have been precisely located in one scan plane before starting the automatic alignment procedure. Due to this initial condition, potential errors related to the z-direction are restricted.

A complete image series of an automatic needle placement trial tilted to the scan plane is shown in figure 8.9. These preliminary experiments provide a kind of technical accuracy of the applied method. In ensuing test series several CT-guided needle placement trials have been performed with different fruits, and finally with a pig cadaver.

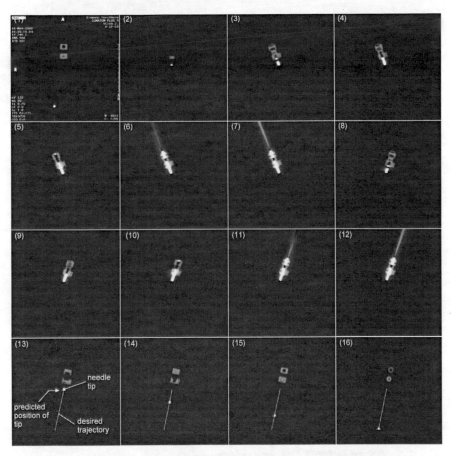

Figure 8.9: A complete image series of an automatic needle placement trial tilted to the scan
plane. During automatic needle insertion the desired needle trajectory and the pre-
dicted tip position is displayed on the image. Thus, the operator can directly see any
deviation between the real tip position and the desired one (small horizontal line
should be located directly below the visible needle tip). The last image (16) shows the
small target bead instead of the needle tip (the needle has been stopped before).

8.2.2 CT-guided Needle Placement Experiments with Fruits

In the following test series in automatic CT-guided needle placement different types of fruits
have been punctured. This provided first information about the influence of penetrating tissues on the needle and its path towards the target. Additionally, it was intended to exchange
the metal target beads with a more realistic target, which allows the penetration of a needle. In
all these trials needle insertion has been performed within the CT scan plane.

Before starting the experiments several types of fruits have been scanned in order to obtain
information about their suitability as test object for the following needle puncture trials. Fig-

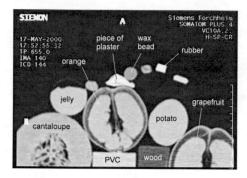

Figure 8.10:   CT image of different fruits and other test materials.

ure 8.10 shows a CT scan with some fruits and other materials. Using a wax bead as target would be advantageous because of the easy plastic deformability under contact with the needle. However, wax does not show very high contrast compared to the fruit flesh. Good absorption for example comes from the piece of plaster (see figure 8.10). After some attempts in creating a suitable target it turned out, that the contrast of the wax could be significantly increased by admixing plaster powder. Thus, the following experiments have been performed

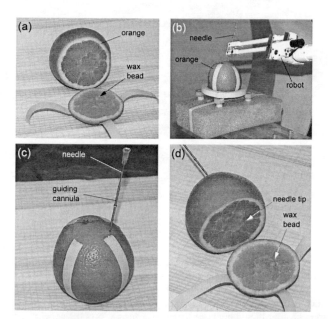

Figure 8.11:   Automatic needle puncture of an orange. (a) The orange is cut and a small wax target bead (Ø6mm) is placed in the fruit flesh. (b) Then the orange is placed on the table simulator and automatic CT-guided needle placement is performed. (c) As soon as the puncture needle is paced in the orange the guiding cannula is detached from the robot. (d) The orange is opened again and the wax bead is removed to determine the puncture mark of the needle in the wax.

by employing small wax beads (Ø6mm) provided with plaster powder.

Experiments have been conducted with several apples, a cantaloupe, and an orange. Figure 8.11 shows the automatic needle puncture of the orange. First, the orange is cut and the wax bead is placed in the fruit flesh before the orange is closed again with two stripes of tape. Then the orange is fastened with a radiolucent fixture on the table simulator before automatic CT-guided needle placement is performed (see figure 8.11-b). As soon as the puncture needle is paced in the orange, the guiding cannula is detached from the robot and the orange is opened again to remove and investigate the punctured wax bead.

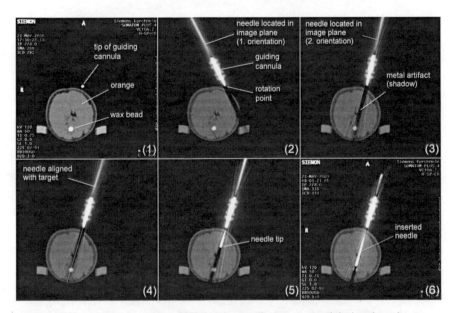

Figure 8.12:   Automatic needle puncture of an orange. The wax target and the insertion point are located in the same scan plane. (1) The guiding cannula's tip is visible in the scan plane as a bright spot. The robot is automatically tilting the guiding cannula into the image for two different needle poses (2)(3). Then, the needle is moved within the scan plane till alignment with the target is achieved (4). Finally, the needle is inserted manually (5)(6).

Figure 8.12 presents several CT scans taken during the automatic needle puncture of the orange. The wax bead and the insertion point are located in the same scan plane. Therefore, the complete procedure can be observed without moving the CT table or the table simulator. The tip of the guiding cannula is visible in the image as a bright spot. Similarly, the wax bead, which has been admixed with plaster powder, shows up with good contrast. After the robot has automatically tilted the guiding cannula into the scan plane for two different needle poses, the scan plane can be determined in the robot coordinate system. Then, the needle is moved within the scan plane till alignment with the target is achieved (compare figure 8.12-4). During these experiments it turned out that the needle drive of the robot is not strong enough to push the needle through tough tissues as the peeling of an orange. Therefore, the needle has been manually advanced into the orange after alignment has been achieved. This manual needle insertion was performed under intermittent scanning till the target bead has been reached (figure 8.12-6).

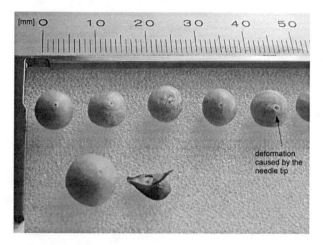

Figure 8.13:   Different targets: several wax beads and an apple seed have been used as target. All of these targets were successfully punctured during CT-guided needle placement experiments and therefore show a small puncture mark caused by the needle tip.

If the needle has successfully hit the target, the surface of the wax bead clearly showed a small puncture mark caused by the needle tip (see figure 8.13). Further needle placement experiments have been performed with a cantaloupe and apples. However, the puncture of a small wax bead or an apple seed give more qualitative information about the precision in needle placement. The deviation of the needle from the ideal path is not precisely definable by examination of the wax beads after puncture (compare figure 8.13). On the other hand, the simple and precise approach of using a CCD camera to determine the remaining deviation vector after needle placement, as presented above, is not possible due to the invisibility of the target embedded in fruit flesh. Therefore, another approach has been followed in order to estimate the remaining deviation after needle placement:

It is assumed that after needle placement the 3D deviation of the needle geometrically consists of two components: (i) a deviation within the CT scan plane, which can be measured in the CT scan itself, and (ii) a deviation perpendicular to the scan plane, not detectable in the CT scan but assessable by the CT slice thickness and the needle length. For example, with a slice thickness of d and a visible needle length L in the image, the needle can be tilted perpendicular to the scan plane for $\pm\arcsin(d/2L)$ without changing the needle's appearance in the image (compare figure 6.17). These consideration demonstrate that the smaller the CT slice thickness is chosen, the higher the geometrical precision of positioning the needle within it. This is the reason why all presented experiments in automatic CT-guided needle placement are performed with the thinnest slice thickness of 1 millimeter, which is achievable with the employed CT scanner.

Both deviation components together lead to an estimate of the maximum possible deviation of the needle in each trial. Since the second deviation component (perpendicular to scan plane) is comparatively small, this approach seems to be acceptable without considerably falsifying the results.

no.	target location	insertion depth	deviation within scan plane (after insertion)	max. deviation by stopping criterion ($\delta = 0.2°$)	max. deviation perpendicular to scan plane	estimated 3D deviation (max.)
1	apple	59 mm	0.38 mm	0.21 mm	0.38 mm	0.54 mm
2	apple	58 mm	0.30 mm	0.20 mm	0.38 mm	0.48 mm
3	apple	41 mm	0.52 mm	0.14 mm	0.28 mm	0.59 mm
4	apple	55 mm	0.75 mm	0.32 mm	0.60 mm	0.95 mm
5	apple	55 mm	0.32 mm	0.30 mm	0.55 mm	0.63 mm
6	apple	70 mm	0.53 mm	0.25 mm	0.46 mm	0.70 mm
7	apple	51 mm	0.28 mm	0.19 mm	0.36 mm	0.45 mm
8	cantaloupe	92 mm	0.74 mm	0.19 mm	0.36 mm	0.82 mm
9	cantaloupe	85 mm	0.36 mm	0.21 mm	0.39 mm	0.53 mm
10	orange	71 mm	0.51 mm	0.18 mm	0.33 mm	0.61 mm
					mean of deviation $|\Delta|$	0.63 mm
					standard deviation	0.16 mm

Table 8.7: CT-guided needle placement experiments with different fruits. In all trials the needle trajectory has been located within the CT scan plane.

Table 8.7 shows the results of the needle puncture experiments in the fruits. The estimated 3D deviation after needle placement is composed of the deviation within the CT scan plane and the estimated maximum deviation perpendicular to the scan plane. Furthermore, the deviation within the scan plane after needle insertion is composed of the deviation caused by the stopping criterion of the visual servoing control loop (compare section 6.4.1), and additional errors caused by needle bending or target motion during insertion. The comparison of both values allows an estimate of the proportion of both error influences.

The mean and maximum deviation in these experiments were 0.63 and 0.95 millimeters. Compared to the mean deviation of 0.26 in the experiments with a freestanding target bead (see table 8.6: placement within the scan plane) the impact of needle bending and drift during insertion is getting obvious.

8.2.3 CT-guided Needle Placement Experiments with Pig Organs

In ensuing test series a pig cadaver has been employed for puncture trials with a more clinical relevance. The cadaver was similar to that used for X-ray guided needle placement described in section 7.4.3. It consisted of different organs, as the kidneys, the liver, the lungs, and the heart. Again, the pig cadaver was put into a transparent plastic bag to prevent its surface from drying up and to facilitate its handling during experiments. Additionally, a piece of the shell of a water melon was placed on the cadaver to simulate the abdominal wall and to stabilize the underlying tissues. Figure 8.14 shows the setup with the plastic bag lying on the radiolucent experiment table of the CT table simulator. A small metal bead with a diameter of 2 millimeters served as target. A thin wire is soldered to the bead (compare figure 7.19), which would allow for electrical contact sensing between the needle and the metallic bead during needle insertion and facilitates handling of the bead within the cadaver.

The problems that occurred during needle placement in the cadaver have been similar to that in X-ray guided needle placement: needle bending and drift during insertion, and the precise control of the needle insertion depth.

Several needle placement test series were performed, while the target has been placed in the cadaver lungs, the liver, or the kidneys. Figure 8.15 shows several CT-fluoroscopy images taken during automatic needle placement in the lungs of the pig cadaver. In this test series the insertion point and the target structure were *located in the same scan plane*. The target bead is

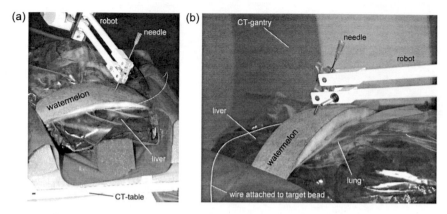

Figure 8.14: In this test series different pig organs are used for needle placement. The photos are taken during CT-guided puncture experiments in the liver.

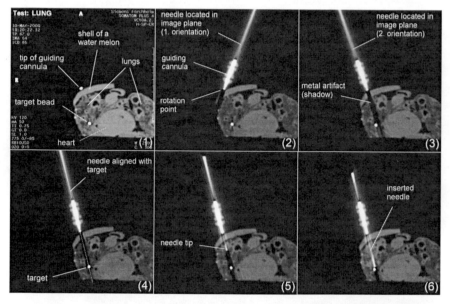

Figure 8.15: CT-scans taken during the automatic needle placement procedure in the lungs. Insertion point and target are located in the same scan plane. A small metal bead (∅ 2mm) serves as target which is placed in the lung of a pig cadaver. (1) The guiding cannula's tip is visible in the scan plane as a bright spot. The robot is automatically tilting the guiding cannula into the scan plane for two different needle poses (2)(3). Then, the needle is moved in the scan plane till alignment with the target is achieved (4). Finally, the needle has been inserted manually (5)(6).

no.	target location	insertion depth	deviation within scan plane (after insertion)	max. deviation by stopping criterion $(\delta = 0.2°)$	max. deviation perpendicular to scan plane	estimated 3D deviation $|\Delta|$ (max.)
1	kidney	67 mm	0.27 mm	0.23 mm	0.44 mm	0.51 mm
2	kidney	62 mm	0.45 mm	0.22 mm	0.40 mm	0.60 mm
3	kidney	67 mm	0.24 mm	0.24 mm	0.44 mm	0.50 mm
4	liver	77 mm	1.68 mm	0.27 mm	0.52 mm	1.76 mm
5	lung	57 mm	0.12 mm	0.20 mm	0.37 mm	0.39 mm
					mean of deviation $|\Delta|$	0.75 mm
					standard deviation	0.57 mm

Table 8.8: CT-guided needle placement experiments with a pig cadaver. In all trials the needle trajectory has been located within the CT scan plane.

visible in the image as a bright round spot. The stepwise process of needle placement is identical to that shown above in the puncture of the orange. After the robot has automatically
tilted the guiding cannula into the scan plane for two different needle poses, the scan plane
can be determined in the robot coordinate system. Then, the needle is moved within the scan
plane till alignment with the target is achieved (compare figure 8.15-4). In these cadaver experiments, the needle drive of the robot is again not strong enough to push the needle through
the tough tissues of the pig organs. Therefore, the needle is manually advanced into the cadaver with intermittent CT-fluoroscopy scanning after alignment is achieved (compare figure
8.15-6).

In these cadaver trials the needle has been inserted within the CT scan plane. The remaining 3D deviation between the needle axis and the target after needle insertion has been estimated similarly to the approach described above: the 3D deviation can be split in a deviation
component within the CT plane (measurable in the CT image), and a second component perpendicular to the scan plane which has been estimated (compare page 146). The results of
these needle placement experiments are presented in table 8.8. The mean of the estimated 3D
deviation is 0.75 millimeters. This is slightly larger than the mean deviation in the CT-guided
needle placement experiments in the fruits (0.63mm). The reason for this might be the more
homogenous structure of the punctured fruit flesh than of the cadaver tissues. If the needle

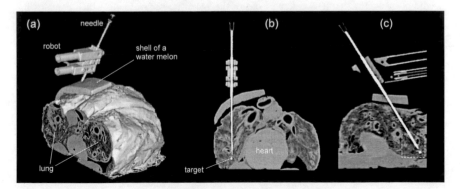

Figure 8.16: Needle placement in a pig's lung. To determine the deviation between the needle axis and
the target after needle insertion, a high resolution spiral scan has been performed (a). Figure (b) and (c) show two MPRs (multi planar reformation) of the 3D volume.

penetrates tough tissues in an angulated orientation, the needle tends to drift away from the straight path. Therefore, a larger mean deviation in the cadaver trials compared to that in the fruit punctures is not surprising.

In further CT-guided needle placement trials with the pig cadaver the needle has been inserted tilted to the scan plane. In order to determine the remaining deviation between the needle axis and the target after needle placement, a high resolution spiral scan of the puncture region has been performed. Figure 8.16-a shows the 3D reconstructed pig cadaver together with the needle and the distal end of the robot. With the help of a 3D image viewing software, slices with arbitrary orientation have been extracted out of the reconstructed 3D volume (see figure 8.16-b,c). These slices, so-called MPRs (multi planar reformation), were used to determine the remaining deviation Δ after needle placement.

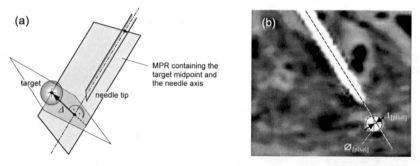

Figure 8.17:   The remaining deviation Δ between the needle axis and the target center is measured with the help of a MPR, which is extracted out of the reconstructed 3D volume of the puncture region. (a) The desired MPR contains the needle axis and the target midpoint. (b) This easily allows to measure the deviation $\Delta_{[pixel]}$ and the diameter of the target bead $\varnothing_{[pixel]}$ in terms of pixels.

Figure 8.17-a demonstrates the principle of how the remaining deviation Δ is determined after needle placement. A MPR is extracted out of the reconstructed volume, which contains the needle axis and the target midpoint. The orientation of the MPR is defined with the help of a 3D viewing software provided with the CT scanner. The deviation $\Delta_{[pixel]}$ and the diameter of the target bead $\varnothing_{[pixel]}$ in the MPR can be easily measured in terms of pixels (see 8.17-b). Since the real diameter of the bead is known in millimeters ($\varnothing = 2mm$), the relation η between millimeter and pixel size could be easily computed: $\eta = \varnothing : \varnothing_{[pixel]}$. Therefore, it is straightforward to compute the resulting deviation Δ in millimeters between the needle axis and the target bead: $\Delta = \eta \cdot \Delta_{[pixel]}$ (η is constant near the target bead). In these experiments the deviation has been determined as a length within the double oblique MPR slice, not as a 3D vector.

Tests in automatic CT-guided needle placement have been performed in the cadaver's lungs, the liver, and in a kidney. The results are shown in table 8.9. The mean of the estimated deviation between the needle axis and the target center is about 1 millimeter. This is slightly larger than the mean deviation in the cadaver punctures when the needle has been placed within the CT scan plane (compare table 8.8). This might be due to the larger insertion depth in the latter experiments.

no.	target location	insertion depth	tilt angle to scan plane	CT-fluoroscopy duration	estimated 3D deviation $|\Delta|$
1	lung	77 mm	32°	39 sec	0.6 mm
2	kidney	72 mm	25°	40 sec	0.7 mm
3	liver	87 mm	22°	40 sec	1.6 mm
				mean of 3D deviation $|\Delta|$	0.97 mm

Table 8.9: CT-guided needle placement experiments with a pig cadaver. A metal bead with a diameter of 2mm served as target. In all trials the needle trajectory was oriented tilted to the CT scan plane.

8.2.4 Experiments with the 'Active Needle'

The principle and the structure of the self-bending needle, called 'active needle', has been presented in section 5.7. The prototype consists of two bent cannulas nested into each other. It allows defined bending of the needle in order to correct occurring needle drift. Additionally, the active needle enables the puncture of targets located behind bones or large vessels and that do not allow for a straight access trajectory.

First puncture experiments with the active needle have been performed under CT imaging with the pig cadaver (see figure 8.18-a). The active needle is hold by the guiding cannula of the needle guiding robot. However, in order to utilize the new degrees of freedom provided by the active needle, it has to be moved and positioned precisely within the robot coordinate frame. This is not possible with the actual configuration of the needle guiding robot. Additionally, definition and execution of the puncture trajectory within the patient's anatomy is getting more challenging for the physician. Therefore, a *preoperative interventional planning tool* would be required for defining the most appropriate access to the target. Thus, the experiments with the active needle have been restricted to simple puncture tests without trying to hit a certain target.

Figure 8.18-b shows a CT scan of the active needle placed in the cadaver's liver. The metallic needle drives of the active needle are located outside the scan plane so that they do not cause any image artifacts. Therefore, the drives are not visible in the image. In this experi-

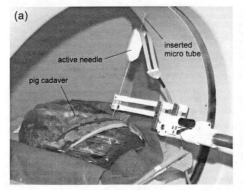
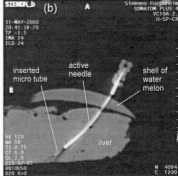

Figure 8.18: Placement of the active needle in the lung of the pig cadaver. The active needle is mounted to the robot's end-effector (a). The CT scan shows the bent needle located within the scan plane (b). A micro tube is put through the active needle.

ment a micro tube (Ø0.6mm) has been inserted through the active needle which can be used e.g. to puncture a target. A 3D reconstruction of a spiral scan of the pig cadaver with inserted active needle is shown in figure 8.19. These images emphasize once more that pre-interventional planning of an oblique and curved access trajectory is essential to define the right insertion point, needle orientation, and bend radius for the active needle.

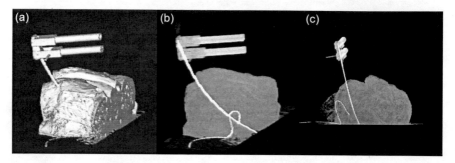

Figure 8.19: Reconstructed 3D volume, scanned after placement of the active needle in the pig's lung. Image (b) and (c) show two MIPs (maximum intensity projection) from different views.

Chapter 9
Discussion of Results

The automatic needle placement experiments presented in the previous two chapters were performed basically in order to gain quantifiable results about the precision that is achievable by the proposed procedures. There are several sources for inaccuracies which have influence on the resulting accuracy of the entire needle placement procedure. But not only the precision achievable in these two approaches is of interest. There are further issues which have to be considered in order to estimate the *user acceptance*. In this chapter the proposed needle placement approaches for X-ray or CT imaging are discussed with regard of following aspects:

- precision
- safety
- radiation exposure
- costs

At the end of this chapter further improvements for the needle guiding robot are suggested, which may empower the system for a clinical use.

9.1 Precision

Doubtless one can say that the achievable needle placement precision is the most important characteristic of an automatic needle placement system. However, many factors contribute to inaccuracies and thus have a bad impact on the overall system accuracy. They can be divided in inaccuracies related to (i) the *automatic needle alignment* process, and (ii) the *needle insertion* process.

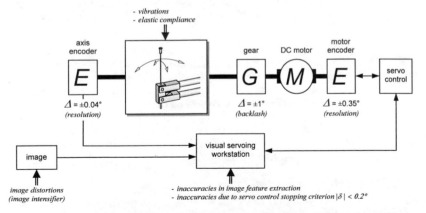

Figure 9.1: Several factors have impact on the accuracy of the *needle alignment* procedure: (a) the transmission chain: gear backlash, motor encoder resolution, axis encoder resolution; (b) static and dynamic disturbance: vibrations, elastic compliance; (c) image distortions and inaccuracies in image feature extraction.

9.1.1 Errors Related to the Automatic Needle Alignment Process

Figure 9.1 shows a flow chart which summarizes the sources and influences of errors on the accuracy of the needle alignment process. For example the transmission chain of the robot is affected by gear backlash and the restricted resolution of the motor- and axis-encoders. Furthermore, static and dynamic disturbances, as vibrations or elastic compliance, are effecting the mechanical stability of the robot. Finally, image distortions (only with X-ray image intensifier), inaccuracies in segmentation and image feature extraction (discrete and blurred object boundaries) and the stopping criterion of the visual servoing feedback loop ($|\delta| < 0.2°$) are leading to inaccuracies in the alignment process.

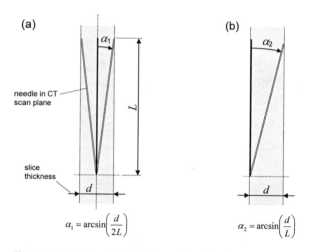

Figure 9.2: Illustration of the deviation angle α, with which the needle can be tilted perpendicular to the scan plane without changing the needle's appearance in the CT image. Angle α depends on the CT slice thickness d and the visible needle length L in the CT image. (for example: d=1mm, L=180mm: α_1=0.16°, α_2=0.32°)

In case of CT-guided needle placement the slice thickness of the CT scan is a crucial factor for the alignment accuracy in the proposed approach. Figure 9.2 illustrates the relationship between the slice thickness d, the visible needle length L in the image, and the deviation angle α, with which the needle can be tilted (perpendicular to the scan plane) without changing its appearance in the CT image. During the alignment procedure, the robot is tilting the needle into the scan plane for several times in order to register the CT scan plane with the robot's coordinate frame. The thinner the CT slice thickness, the smaller the deviation angle, and the higher the accuracy in the needle alignment process. Figure 9.2-b shows the setting with the largest possible deviation angle of $\alpha = \arcsin(d/L)$.

9.1.2 Errors Related to the Needle Insertion Process

Even in case that the needle would be perfectly aligned with the target, the needle insertion process itself may cause significant deviations due to needle bending and needle drift during insertion.

   In case of CT-guided needle insertion within the scan plane this can be a major problem, because the needle may leave the CT scan plane unintentionally. This leads to the fact that the

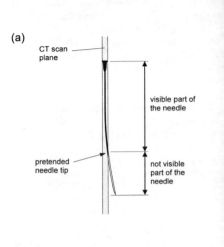

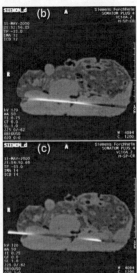

Figure 9.3: Needle bending can be a major problem in precise CT-guided needle placement, if the needle is leaving the CT scan plane unintentionally (a). The pretended needle tip is the last visible part of the needle before it is leaving the
scan plane. The CT images (b) and (c) show two slices of a bent needle in the
pig cadaver.

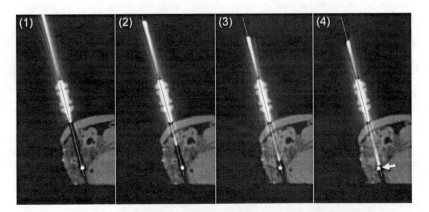

Figure 9.4: Needle placement within the lung of the pig cadaver. Close before the needle tip reaches the target, the bead is drifting away (3). The reason for this
is the deformation of the lung tissue close to the needle tip which has influence on the embedded target bead as well (arrow).

robotic system or the operator has no more visual control of the needle tip or the actual insertion depth. It may happen, that although the needle is advanced into the body, the CT image shows no movement of the pretended needle tip. Figure 9.3 illustrates the reason for this effect. Actually, only a part of the needle is visible in the CT image. The pretended needle tip is the last visible part of the needle before it is leaving the scan plane. Therefore, the tip seems not to move during insertion. Increasing the slice thickness would reduce this problem but lowers the needle placement accuracy at the same time (compare figure 9.2).

The major reason for needle bending is the asymmetric grinded needle tip. In order to reduce this effect, all cadaver experiments have been performed with a needle provided with a sharp symmetric tip (compare section 7.4.3).

A further problem in needle placement is the movement and deformation of soft tissues under the impact of the advanced needle. This may lead to a drift of the target structure away from its initial location. Figure 9.4 shows an example of this effect in case of needle placement within the lung of the pig cadaver.

		freestanding target			target in pig cadaver		
		mean deviation	max. deviation	mean insertion depth	mean deviation	max. deviation	mean insertion depth
X-ray guided	CCD imaging	0.2 mm	± 0.4 mm	(57 mm)	—	—	—
	X-ray imaging	0.4 mm	± 0.7 mm	78 mm	1.25 mm	± 1.5 mm	80 mm
CT guided	within scan plane	0.26 mm	± 0.4 mm	80 mm	0.75 mm	± 1.0 mm	66 mm
	tilted to scan plane	0.6 mm	± 1.0 mm	83 mm	1.0 mm	± 1.6 mm	79 mm

Table 9.1: Overall accuracy in the X-ray- and CT-guided needle placement experiments. The table shows the mean and max. deviations between needle axis and target bead after needle placement.

Table 9.1 summarizes the results of the achieved needle placement accuracy with X-ray- and CT-imaging. It shows the remaining deviations after needle placement (i) with a *freestanding target bead*, and (ii) with a *pig cadaver*. The results with the freestanding bead are not affected by errors related to the needle insertion process. Therefore, a comparison of both test series (freestanding target versus target in pig cadaver) allows to estimate the influence of the needle insertion process on the overall accuracy. This comparison points out that for both imaging modalities the deviations in the pig cadaver trials are about twice as large as the deviations with the freestanding target bead.

However, real needle placement procedures in living patients are affected by further sources of inaccuracies: tissue motions caused by e.g. patient breathing or the beating heart. These conditions, which significantly complicate needle placement, cannot be easily simulated in a laboratory setup. However, under real conditions the impact of patient breathing can be minimized for instance if the needle placement procedure can be performed during one breathhold, or the needle is advanced intermittently in the same respiratory phase.

9.2 Safety

The use of an active robot device to position a passive instrument requires very safe, stable and compliant operation of the manipulator and its control system. Since medical robots are working in direct contact with the patient and the physician, safety considerations are significant in the development process of the robot. Safety considerations concern several issues [184][42]:

- can the manipulator get sterilized ?
- does the robot meet electrical safety requirements ?
- limitations of the system (tolerances and inaccuracies) ?
- safety of robot control software ?
- system behavior in case of failure ?
- degree of autonomy of the system's task ?

Although, the purpose of the initial design of the novel needle guiding robot was not the usage on patients in a clinical surrounding, there have been already different safety features integrated in the system. For instance, to achieve sterile conditions surgical drape can be easily used for the whole robot except the guiding cannula, which is easily detachable and sterilizable separately. Only in case of using the integrated needle drive for remotely controlled needle insertion, sterilization might be difficult. Additionally, the design of the manipulator with its integrated cables allow for cleaning and the use of surgical drape.

Furthermore, the utilization of small 12V drives for robot motions and control signals with 5V does not lead to any electrical hazard.

The robots accuracy in needle positioning is ±0.04° (compare section 5.5), which has no significance for the overall accuracy of the needle placement process (see table 9.1).

A crucial safety feature is the ability for the user to monitor what the system is doing, confirm decisions, and intervene if necessary [184]. In this sense, the developed user interface for planning and controlling the needle placement procedure allows the physician to observe the advancement of the needle in real-time on the monitor. If the deviations of the needle are getting too large on the way to the target the user can abort the procedure immediately. If the needle is inserted manually after automatic needle alignment, the robot switches from computer controlled to motion-disabled mode and the responsibility for needle insertion is put back on the physician.

Another safety feature is the mechanical design of the robot with its simple parallel kinematic. In contrast to robots provided with many joints in order to reach a large work envelope with point-to-point motions, the movements provided by the needle guiding robot are simple rotations of the needle around the insertion point (remote center of motion) obtained by only two degrees of freedom. This reduction of manipulator complexity makes the whole system very much reliable and save.

9.3 Radiation Exposure

Radiation exposure for the physician and the patient is a very important issue in interventional radiology. The following two paragraphs discuss the use of the proposed needle guiding robot in order to reduce X-ray exposure for the physician and the patient:

X-ray exposure for the physician

Interventional radiology requires the operator and assisting personnel to stay close to the patient. These procedures typically require placement of the hands within the radiation field while placing a needle or advancing a catheter under direct imaging. Very often this leads to significant hand doses for the operator [77].

In this context, the use of a robotic tool for conducting simple tasks as needle alignment or needle insertion under X-ray imaging can completely avoid X-ray exposure the operator's hands. The robot can be used remotely controlled (as proposed for needle insertion) or automatically image-guided as demonstrated with the visual servoing needle alignment approach.

X-ray exposure for the patient

While interventional radiological procedures are minimally invasive and beneficial to patients, some complex and lengthy fluoroscopically guided procedures may cause radiation-induced skin injuries [10]. It is difficult to access in the actual state, if the use of a needle guiding robot will really shorten the procedure and reduce radiation exposure for the patient. This would require a clinical validation. However, in the author's opinion there is indication for a reduced radiation exposure for the patient:

Image analysis and feature detection in the acquired images can be done faster by a computer than by a human (assuming that the feature is visible and detectable in the image). Furthermore, the robot can determine quickly and geometrically precise angles and distances in the image. The robot can hereby determine needle orientations in the robot coordinate frame, which is leading to a registration of the image coordinates to the world coordinate system. Finally, the system is capable to place the needle automatically in the desired orientation and keep this pose during manual or remotely controlled insertion.

These considerations are much likely that the needle placement procedure can be performed with a smaller number of acquired images if performed by the needle guiding robot than by a human operator, assuming the same precision in outcome. A smaller number of X-ray images or CT images automatically leads to reduced X-ray dose for the patient.

9.4 Cost-benefit Analysis

The factor 'cost' is essential for the purchase of a robotic system, or more precisely: the cost-benefit analysis. The expense for investment and operation of a robotic system has to be compared with the benefits and potential cost savings associated with it. This section will discuss these aspects with regard to the presented needle guiding robot.

Purchase costs

The needle guiding robot is a very small and quite simple manipulator compared to e.g. the ROBODOC™ system for orthopedic surgery (compare section 3.2). While ROBODOC™ costs about US$ 500.000, the expense for the presented needle guiding robot would be much less. Referring to a radiologist, which has been interviewed by the author about the need of a needle guiding robot on the healthcare market, a total system price of € 25.000 would be acceptable (compare section 4.2.3).

The accrued *material costs* for the entire system (prototype) were approximately € 4.800: about € 1.000 for the manipulator hardware and the passive arm, and another € 3.800 for the control PC with all the plug-in boards. A meaningful estimation of the *development costs* (hardware and software) and the *manufacturing costs* of the needle guiding robot is challenging and has to be derived from the required amount of time:

The expenditure of time for the development and designing work for the manipulator and the passive arm was about 4 man-months. About 2 man-months would be needed for professional manufacturing of the manipulator. The programming of the control software and the user interface for the actual prototype required about 4 man-months for the X-ray setup, and another 3 man-months for the CT-setup. For functional testing of the complete system another man-month should be assessed.

However, if the actual robot should be offered on the health care market, surely a reverse engineering of the entire system would be required with a more sophisticated and reliable control software, user interface, and manipulator hardware than realized for the actual prototype.

Benefits due to employment of the robot

A proper evaluation of the cost-effectiveness of employing this image-guided robot requires consideration of following questions:

- is there a benefit for the patient by employment of the robot ?
- is there a benefit for the physician/medical provider to use the robot ?

The answer to these questions is quite speculative and surly requires further investigation of these issues. For the patient, there would be a benefit if needle placement can be performed with less radiation exposure, less trauma, and less risk due to higher accuracy. The results in robot guided needle placement, presented in chapter 7 and 8, are very promising and let assume that the visually controlled needle-guiding robot allows to perform needle placement with increased precision and speed than in manual techniques.

With its very compact and light design, the robot is appropriate and optimized for usage inside a CT gantry as well as in conjunction with an X-ray C-arm. With the automatic image guided needle placement approach, the robot basically allows to avoid radiation exposure for the operator. Additionally, there is indication that using the robot for demanding needle placement procedures, procedure time for the puncture may be reduced due to better control performance of the needle and increased precision. Therefore, radiation exposure may be reduced even for the patient. Besides the prevention of radiation exposure, there exist further potential benefits for the physician and the medical provider: increased productivity and patient throughput, due to reduced procedure time, by increased quality of care.

However, as soon as the procedure is not very demanding and precision in needle placement is not significant, robot guided needle placement would be most likely not faster than manual performance of the puncture. This may be true especially for highly experienced physicians. The robotic system will not be accepted if it is too expensive for performing only simple applications. But, whenever precision is required in certain needle placement procedures, the robot has certain advantages over a manual execution. For instance, in case of needle placement tilted to the CT scan plane, the proposed approach provides suitable control performance of the needle for improved targeting.

Certainly, further clinical investigations have to be performed to demonstrate that employing this robotic tool allows to increases the quality of the intervention by reduced inhospital costs.

9.5 Suggested Improvements and Further Investigation

The needle guiding system presented in this thesis is still in the early stages of development. Several issues remain to be addressed before this system will be clinically viable. Improvements might be related to the actual manipulator design and its control. However, the objective in this robot development was to design and manufacture a prototype, which allows to validate the novel visual servoing approach for automatic needle alignment in a *laboratory test bed*. But finally, it is aspired to test the robot under real clinical conditions. In order to empower the robot for *clinical use* further improvements have to be implemented. The most significant ones are discussed in this section.

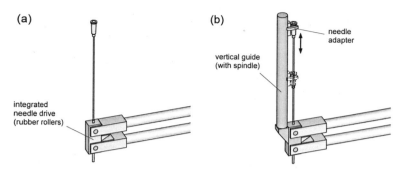

Figure 9.5: Example for a modified needle drive. (a) The actual needle drive with two rubber rollers integrated in the end-effector of the robot. The needle is grasped close to the insertion point. (b) Alternative needle drive with a vertical guide. Here, the distal end of the needle is fixed to a needle adapter which can vertically move.

Manipulator improvements

During the construction and initial testing of the robot, it was noted where redesigning may be required in future versions of the robot. For instance, an improvement would be a modified needle drive for the manipulator. In the current version the needle is grasped between two rubber rollers and driven only by friction. This principle of needle drive can apply only moderate axial forces during needle insertion and does not allow to penetrate tough tissues. This is the reason why all presented needle placement trials within the pig cadaver have been performed with manual needle advancement towards the target after automatic alignment.

In this context, a needle drive would be advantageous which could apply higher forces to the needle. Figure 9.5-b shows an example of a modified needle drive with a vertical guide mounted to the robot's end-effector. The distal end of the needle is fixed to a needle adapter, which can move vertically due to an integrated spindle. A drawback of this design is its bulky end-effector, especially when working inside a CT scanner. For example, in case of CT-guided needle placement in a stout patient, there is not much space left for needle manipulation between patient and gantry cover inside the CT scanner. Therefore, the radiologist may choose the shortest needle possible to puncture the target. With the actual needle drive the robot needs only as much space for needle manipulation as the needle requires. However, the alternative design needs always the same space due to the vertical guide, independent of the length of the chosen needle.

Another issue is the detachment of the needle from the robot after successful needle placement. This may be necessary especially if the target is located close to the lung. Normally, the needle is inserted during a breathhold so that no tissue movement occurs. However, as soon as the target is reached and the patient starts breathing again, the needle may be exposed to large tissue movement. Therefore, it is advisable to unlock the needle from the robot as soon as the needle is placed. The presented manipulator allows to exchange the needle drive with a passive needle holder. In this case, the guiding cannula can be easily detached from the robot's end-effector and allows to remove the manipulator while the needle remains in place. This unlocking of the needle does not work together with the realized needle drive.

Furthermore, it may be advantageous if the needle could automatically detach itself in case that too large lateral forces are applied to it. This could happen e.g. in case of an emergency or if the patient unintentionally moves the body. A technical solution might be the integration of a force- or torque-limiting clutch between the guiding cannula and the robot.

Software improvements

Besides these suggested improvements of the robot's hardware, there are further issues concerning the control software. For instance, there is currently only a simple threshold approach used for detection of the needle in the image. This has been appropriate in the presented experiments under predefined laboratory conditions. For clinical routine a more reliable needle detection algorithm has to be implemented, which applies for example automatic edge detection algorithms.

Furthermore, additional safety features have to be integrated in the control software. For instance, an automatic halt in case that the needle deviation between actual and desired needle path exceeds a certain value during insertion. In addition, the integration of self-test routines into the control software would lead to a more reliable system and could indicate the operator impeding system failures.

The presented experiments using CT-fluoroscopy imaging allows for further improvements concerning X-ray radiation dose for the patient. CT imaging has been performed continuously during the automatic alignment process of the needle. For clinical use with patients it is more appropriate to employ intermittent CT scanning for the visual control of the needle. In conjunction with further optimization of the control loop, this could result in significantly shorter CT-fluoroscopy scan time and radiation dose.

A further improvement in CT-guided needle placement is associated with the type of CT scanner. While in the presented experiments a so-called single slice CT scanner (Somatom Plus4, Siemens AG, Germany) has been employed, newer high performance scanners as the Siemens Somatom Sensation64 provide fast image acquisition with 64 image slices per rotation (multi-slice CT scanner). The use of such a multi-slice scanner for image-guided needle placement has significant advantages for needle control: the middle slice is the desired slice to position the needle within. If the needle is leaving this slice due to needle shift during insertion, its position can be precisely determined in the images of the adjacent slices before and behind. Additionally, this 'volume scanning' provides better control performance even in an orthogonal needle approach. However, the implemented visual servoing algorithms would have to be adapted for multi-slice CT scanning.

9.6 Subsequent Work of other Research Groups

The author's first publication in February 2000 on the new principle on visual servoing under X-ray fluoroscopy [1], and its follow-up paper in October 2000 using CT-imaging [2], have generated a lot of interest in the scientific community. In the following two paragraphs some recent research work in this new field is presented.

Visual servoing using X-ray fluoroscopy imaging

In October 2000, Patriciu et al. [119] presented the same visual servoing technique using X-ray fluoroscopy imaging, which they intelligently named fluoro-servoing. They used the RCM-PAKY robot [154] for visually controlled needle alignment. The basic difference is, that Patriciu started the automatic alignment process by moving the needle stepwise around the insertion point on a "conic surface" in order to determine a rough result for the viewing direction of the X-ray C-arm. This makes the alignment process more independent from starting conditions and ensures, that the servo-plane in which the needle is moved (plane π_1 in figure 6.4) is perpendicular to the viewing direction. However, this automatic servoing applies additional X-ray exposure to the patient. In the approach presented in this thesis (see section 6.3.1, Step 1) it is the operator, who is responsible for choosing a servo-plane that is about perpendicular to the viewing direction of the X-ray C-arm. Patriciu *et al.* also did preliminary testing of the approach with a video-camera based mockup using a black needle in front of a white background. The results are quite similar to that presented in section 7.5.3 and 7.5.4 of this thesis [121].

Visual servoing using CT imaging

The author's work on visual servoing using CT-imaging [2] inspired new research in this field. For example, in October 2001 Patriciu *et al.* [120] proposed to register their robot (RCM-PAKY [154]) to the CT image space with the help of the laser markers readily available on any CT scanner. Commonly, two perpendicular laser fan-beams are projecting a cross onto the patient in a defined geometrical relation to the scan plane. The proposed registration method is based on a stepwise placement of the robots needle within these two laser planes. This is done manually by the operator via joystick control, while the desired needle orientation within the laser plane is as well determined and adjusted by the operator via joystick. After positioning the needle in two dissimilar orientations within the 'image laser plane', the plane is defined in the robots coordinate system.

This approach is an interesting version of the automatic registration method presented in section 6.4. The basic difference is that Patriciu proposes the registration procedure being done with respect to the CT laser marker system instead of the CT image space. Advantageous in Patriciu's approach is that this requires less CT-images during registration and thus causes less radiation exposure for the patient. However, shortcomings of the approach are due to the manual (joystick) and visual control of the needle done by a human operator during the registration procedure. This non-automatic approach will not only make the procedure lengthy but may lead to inherent inaccuracies as well.

Other researchers have continued the presented approach and have referenced some of this work done in this thesis. For example, Masamune *et al.* [102] and Fichtinger *et al.* [50], who have further developed the approach of Susil *et al.* [156] (compare section 2.3.4). Their robot (RCM-PAKY [154]) has been successfully employed for CT-guided percutaneous kidney procedures [155] and prostate biopsy.

A research group at the Georgetown University Hospital, Washington, DC, USA (Cleary *et al.* [30][31]), developed a similar robotic setup in combination with a CT scanner, as presented in this thesis. The setup consists of (a) a mobile CT scanner, (b) the RCM-PAKY robot mounted on the CT table, and (c) an optical navigation system (Hybrid Polaris, Northern Digital, Waterloo, Canada). This setup has been developed for percutaneous spinal procedures, as biopsies for example. Planning and execution of the needle puncture is based on preoperatively acquired CT images. During the intervention the robot has to be registrated first to the spinal anatomy of the patient via navigation system. Then, the manipulator automatically orients the needle according to the preoperatively defined needle access path. Although, this workflow represents a typical CAS procedure without intraoperative real-time image feedback or a visual servoing control of the robot, this group defined the long-term goal of their system as 'integrated robotic system directly linked to X-ray fluoroscopy and CT' [32][36].

Chapter 10

Summary and Concluding Comments

10.1 Summary

Advances in several technologies within the last decades have resulted in substantial changes in medicine, especially in the field of radiology and surgery. Doubtless, milestones have been the introduction of CT and MRI scanners into clinical routine in the 1970s and 80s, which provided high-resolution, cross-sectional images with real 3D information for the first time. This novel capability of sophisticated 3D imaging had not only great impact on the quality of diagnostics. Accurate 3D information of the patient's anatomy also enabled precise planning of the intervention. Furthermore, the integration of computer assistance in the OR on basis of *preoperative 3D image data* in combination with *stereotactic systems*, allowed for guiding a therapeutic tool during surgery. Although, these computer assisted surgeries (CAS) provide improved accuracy with which a surgical procedure can be performed, they have a significant drawback: CAS relies on preoperatively acquired image data for guiding a surgical tool, and assumes that there are no changes in the anatomical structures after image acquisition and during the surgery. However, almost all surgical procedures entail movement of tissue, through the use of retractors, by surgical resection or due to the drainage of fluids.

An answer to this problem is the use of *intraoperative real-time image guidance* to control a therapeutic tool during surgery. Intraoperative imaging provides permanently updates about the patient's anatomy and allows the operator to observe e.g. moving organs or the advancement of a surgical instrument inside the patient's body. This type of interactive visualization is sufficient to guide during most percutaneous biopsies and intravascular interventional procedures.

Appropriate imaging modalities for intraoperative imaging are for example mobile X-ray fluoroscopes. But also CT-scanners are increasingly employed for image-guided interventional procedures or even during surgeries. However, a basic disadvantage of image guided interventions using X-ray fluoroscopy and CT-imaging is the *radiation exposure* for the operator, especially if the intervention is performed under direct imaging.

A solution to this problem has been presented in this thesis: in order to avoid radiation exposure to the surgeon, a *robotic manipulator* has been proposed for remotely or even automatic image guidance of a surgical tool. The operator can stand away from the X-ray beam or may even leave the operating room during the fluoroscopic or CT-guided intervention.

A special *image based control* (visual servoing) has been developed and implemented for the robot, which allows semi-automatic and uncalibrated alignment of a surgical tool under real-time image guidance: real-time images are analyzed and directly taken to achieve surgical tasks with the manipulator. A distinct advantage of this real-time visual control is - over common CAS procedures - that *no prior calibration or registration* of the robot or the imaging modality is required. This means that no additional stereotactic systems are needed.

An uncomplex but challenging minimally invasive task is the precise placement of a puncture needle, e.g. for percutaneous biopsy, drainage, or tumor ablation. Needle placement is commonly performed manually and, especially in high risk areas, under direct image guidance. Therefore, *needle placement* has been identified as the most suitable clinical application for testing of the novel robotic manipulator and its visual servoing control. This thesis demonstrated the efficacy and precision of this new approach by several phantom and cadaver studies. Furthermore, to the author's knowledge, this work presents the first implementation and experimental results on visual servoing using X-ray-fluoroscopy or CT-imaging.

Chapter 1 presented the basic principle of the novel interventional imaging workplace consisting of the image controlled micro-robot and the X-ray imaging modalities. The following chapter 2 gave an overview of the main issues in computer assisted surgery (CAS) and discussed several manual needle placement techniques currently in clinical use. Furthermore, several approaches of different research groups have been presented, where *needle guiding robots* are employed under X-ray or CT-imaging guidance.

Chapter 3 presented a brief survey of different types of robotic systems currently used in the medical field. These systems are applied to neurosurgery, orthopedics, and minimally invasive surgery.

The usefulness of a robotic manipulator for image-guided needle placement has been investigated in a market analysis conducted by the author, which has been presented in chapter 4. Eighteen physicians of different medical faculties were interviewed about potential key-applications for the proposed visually controlled robot. The resulting key application has been defined as: *automatic and visually controlled placement of a puncture needle under X-ray or CT-imaging.* Furthermore, the discussions with the interviewees allowed to derive clinical requirements for the robotic manipulator.

The development of the robot has been presented in chapter 5. A simple and inexpensive design has been realized. Due to its compact design, the robot is appropriate for usage inside a CT-scanner or in combination with an X-ray C-arm. Furthermore, the mechanical stiffness and manipulator accuracy has been verified, and the kinematic model of the manipulator has been derived.

Essential in this new approach is the employment of a visual control technique for the robot (*visual servoing*). The theoretical framework of this visual control using X-ray- or CT-imaging has been presented in chapter 6. A distinct advantage of this approach is that no prior calibration or registration of the robot or the imaging system is needed. All required software and hardware components are integrated on the PC-based *visual servoing workstation*, which has been presented in the second part of chapter 6. The software comprises of the user interface and the underlying software components for image acquisition, image analysis, and control of the manipulator.

Chapter 7 introduces the experimental setup and describes the needle placement experiments under *X-ray imaging* performed at the Johns Hopkins University and the Siemens Medical Solutions laboratories. In preliminary experiments a CCD-camera was employed for image acquisition. The test conditions have been arranged in a way that the acquired images have been very similar to the real X-ray images. This allowed to implement and validate the X-ray visual servoing test bed very conveniently under lab conditions without having to deal with radiation exposure. In ensuing experiments a mobile X-ray C-arm has been employed for imaging. After successful testing of the visual servoing control with a simple, freestanding target bead, a number of cadaver trials were performed in order to achieve more realistic, clinical conditions for the needle puncture experiments. The mean deviation in the X-ray guided cadaver trials has been 1.25 millimeters (needle insertion depth 60..90 millimeters).

Experiments in automatic needle placement using *CT-imaging* have been presented in chapter 8. All CT-guided puncture trials were performed with a Somatom Plus4 single slice CT scanner (Siemens AG, Germany) at the Siemens CT laboratories in Forchheim, Germany. Since the workflow for automatic needle placement with CT imaging is different to that with X-ray imaging, the user interface and the visual servoing control software have been slightly modified. Initial experiments with a freestanding metal target bead were performed to verify the principle accuracy of this new CT-guided approach. Ensuing experiments were conducted with several fruits and finally with a pig cadaver. The mean deviation in the CT-guided cadaver experiments has been about 1 millimeter (needle insertion depth 60..90 millimeters).

The experimental results were discussed in chapter 9. The sources for inaccuracies which had influence on the resulting accuracy of the entire needle placement procedure have been considered as well as further issues affecting the user acceptance: safety, the reduction of radiation exposure, and costs.

10.2 Concluding Comments

In recent years technological advances have resulted in substantial changes in medicine. Different areas of research have arisen developing new technologies of imaging, 3D visualization, virtual reality, teleoperation, frameless stereotaxis, or even medical robotics. The possibility of using robots in the medical field has attracted much interest. Robots designed for surgery have been shown to have several areas where they perform better than humans [20]. Their accurate spatial awareness is used e.g. in neurosurgery with motorized stereotaxy. Precision, repeatability and reliability are important qualities of robotic devices in surgical applications.

However, since the first successful experiments in employing a robot for surgical tasks in the 1980s, the use of robotic assistance in the medical field is still in the very beginning. During the 1990s several surgical robots got commercially available on the health care market, but the benefit and appreciation of these systems is still much discussed among both medical experts and the medical engineering industry. The decision of a medical facility to invest in a surgical robot strongly depends on the benefits and cost-savings entailed with the system. For instance, a new robot has to demonstrate improved surgical accuracy, which may result in improved clinical outcomes, less invasive surgical procedures, faster recuperation and thus shorter hospitalization and reduced long-term costs. But the prove, if robot supported interventions are of higher quality and thus will save costs in a long-term is still not demonstrated for many medical applications [147][41].

The image-guided micro-robot, which has been presented in this work, is a straightforward and inexpensive robotic tool for automatic needle placement. The very compact manipulator basically consists of simple commodity components (about €1000 material costs) and is easily assembled. The robot works under direct supervision of the operator, with the purpose to support the physician in performing needle placement procedures with improved precision. In addition to increased accuracy, the system potentially accelerates the procedure and reduces the X-ray exposure for both surgeon and patient.

Surely, the current prototype is not appropriate for clinical use yet and requires certain improvements, as e.g. (a) sophisticated algorithms for reliable needle detection in the image, (b) fast unlocking of the needle after placement, or (c) the integration of further safety features. However, the current status of development and the presented results are very promising that the system has the potential to fulfill the requirements for clinical use without getting complex, bulky, and expensive as many other medical robotic systems. With its precise image-based control and the optimized design for X-ray guided interventions, the presented micro-robot may be a further step into sophisticated but cost-effective surgical treatment.

References

Publications by the author:

[1] M. H. Loser, N. Navab, B. Bascle, R. H. Taylor, "Visual servoing for automatic and uncalibrated percutaneous procedures", in Proc. *SPIE Medical Imaging 2000: Image Display and Visualization*, S. K. Mun (Ed.), vol. 3976, pp. 270-281, Feb. 2000.

[2] M. H. Loser, N. Navab, "A new robotic system for visually controlled percutaneous interventions under CT fluoroscopy", in Proc. *Medical Image Computing and Computer Assisted Intervention, MICCAI'00*, pp. 887-896, Oct. 2000.

[3] N. Navab, B. Bascle, M. Loser, B. Geiger. R. H. Taylor, "Visual servoing for automatic and uncalibrated needle placement for percutaneous procedures", in Proc. *IEEE Int'l. Conf. on Computer Vision & Pattern Recognition , CVPR 2000*, vol. 2, pp. 327 –334, 2000.

[4] B. Bascle, N. Navab, M. H. Loser, B. Geiger, R. H. Taylor, "Needle placement under X-ray fluoroscopy using perspective invariants", in Proc. *IEEE Workshop on Mathematical Methods in Biomedical Image Analysis, MMBIA-2000*, pp. 46-53, 2000.

[5] M. H. Loser, "Entwicklung mikrotechnisch gefertigter Operationsinstrumente aus Nickel-Titan für die Neurochirurgie", diploma thesis, *University of Karlsruhe*, FZK, 1996.

[6] E. Guber, N. Giordano, A. Schüssler, O. Baldinus, M. H. Loser, P. Wieneke, "Nitinol-based microsinstruments for endoscopic neurosurgery", in Proc. *Actuator 96: 5th Intern. Conf. on New Actuators*, Ed. H. Borgmann, pp. 375-378, 1996.

[7] M. H. Loser, "Mediziner-Befragung - Thema: Durchleuchtungsgesteuertes Führungssystem", *Technical Report:* Siemens AG, Medical Engineering, R&D Department, Erlangen, Germany, 1998.

References:

[8] J. Angeles, *Fundamentals of robotic mechanical systems: theory, methods, and algorithms*, Mechanical Engineering Series, Springer-Verlag New York Inc., New York, USA, 1997.

[9] H. Anton, *Elementary Linear Algebra*, 8th edition, John Wiley & Sons Inc., New York, 2000.

[10] B. R. Archer, " High-dose fluoroscopy: the administrator's responsibilities", in J. *Radiology Management*, vol. 24(2), pp. 26-32; 2002.

[11] L. Axel, "Simple method for performing oblique CT-guided needle biopsies", in *Annual Journal of Radiology*, vol. 143, p. 341, 1984.

[12] A. Bab-Hadiashar, *Data segmentation and model selection for computer vision*, Springer, New York, 2000.

[13] Z. L. Barbaric, T. Hall, S. T. Cochran, D. R. Heitz, R. A. Schwartz, R. M. Krasny, M. W. Deseran, "Percutaneous nephrostomy: Placement under CT and fluoroscopy guidance", in *American J. of Roentgenology*, vol. 169, pp. 151-155, 1997.

[14] C. Barbe, J. Troccaz, B. Mazier, S. Lavallée, "Using 2.5D echography in computer assisted spine surgery", in *Proc. IEEE Engineering in Med. and Biology*, San Diego, pp. 160-161, 1993.

[15] F. Betting, J. Feldman, N. Ayache, F. Devernay, "A new framework for fusing stereo images with volumetric medical images", in *Proc. VCRMed '95*, Springer Verlag, pp. 30-39, 1995.

[16] M. Börner, A. Lahmer, A. Bauer, U. Stier, "Experiences with the ROBODOC system in more than 1000 cases", in *Proc. Computer Assisted Radiology and Surgery*, H. U. Lemke et al. (Ed.), pp. 689-693, 1998.

[17] K. Bouazza-Marouf, I. Browbank, J. R. Hewit, "Robotic-assisted internal fixation of femoral fractures", in *Journal of Engineering in Medicine*, vol. 209, pp. 51-58, 1995.

[18] R. D. Bucholz, "Advances in computer aided surgery", in *CAR'98*, H. U. Lemke et al. (Ed.), Elsevier Science, pp. 577-582, 1998.

[19] R. D. Bucholz, K. A. Laycock, "Coupling information to surgery through imaging", in *Medical Imaging 2000: Image Display and Visualization*, S. K. Mun (Ed.), Proceedings of SPIE, vol. 3976, pp. 2-9, 2000.

[20] R. A. Buckingham, R. O. Buckingham, "Robots in operating theatres", in *British Medical Journal*, vol. 311, pp. 1479-1482, 1995.

[21] G. F. Buess, M. O. Schurr, S. C. Fischer, "Robotics and allied technologies in endoscopic surgery", in *Archives of Surgery*, vol. 135(2), pp. 229-235, 2000.

[22] C. W. Burckhardt, P. Flury, D. Glauser, "Stereotactic Brain Surgery", in *IEEE Engineering in Medicine and Biology*, vol. 14(3), pp. 314-317, 1995.

[23]   G. C. Burdea, "General-purpose robotic manipulators", in *Computer-Integrated Surgery - Technologie and Clinical Applications*, R.H. Taylor et al. (Ed.), The MIT Press, Cambridge, USA, pp. 257-262, 1996.

[24]   A. Bzostek, S. Schreiner, A. C. Barnes, J. A. Cadeddu, W. W. Roberts, J. H. Anderson, R. H. Taylor, L. Kavoussi, "An automated system for precise percutaneous access of the renal collecting system", in Proc. *1st Joint Conf. of CVRMed and MRCAS*, Grenoble, France, pp. 299-308, 1997.

[25]   J. A. Cadeddu, A. Bzostek, S. Schreiner, A. C. Barnes, W. W. Roberts, J. H. Anderson, R. H. Taylor, L. R. Kavoussi, "A robotic system for percutaneous renal access", in *Journal of Urology*, vol. 158, pp. 1589-1593, 1997.

[26]   C. Canudas de Wit, B. Siciliano, G. Bastin, *Theory of robot control*, Springer-Verlag London Ltd., Great Britain, 1996.

[27]   J. F. Cardella, A. T. Young, D. W. Hunter, W. R. Castaneda-Zuniga, K. Amplatz, "New universal radiolucent handle", in *Radiology*, vol. 155, p. 531, 1985.

[28]   A. Casals, J. Amat, D. Prats, E. Laporte, "Vision guided robotic system for laparoscopic surgery", in Proc. *Int. Conf. on Advanced Robots*, pp. 33-36, 1995.

[29]   W. R. Chitwood, "Endoscopic robotic coronary surgery - is this reality or fantasy ?", in *J. Thoracic and Cardiovascular Surgery*, vol. 118(1), pp. 1-3, 1999.

[30]   K. R. Cleary, D. Stoianovici, N. D. Glossop, K. A. Gary, S. Onda, R. Cody, D. Lindsich, A. Stanimir, D. Mazilu, et al., "CT-directed robotic biopsy testbed: motivation and concept", in Proc. *SPIE Medical Imaging 2001: Visualization, Display, and Image-Guided Procedures*, vol. 4319, pp. 231-236, 2001.

[31]   K. Cleary, M. Clifford, D. Stoianovici, M. Freeman, S. K. Mun, V. Watson, "Technology improvements for image-guided and minimally invasive spine procedures", in *IEEE Transactions on Information Technology in Biomedicine*, vol. 6(4), pp. 249-261, 2002.

[32]   K. Cleary, D. Stoianovici, A. Patriciu, D. Mazilu, D. Lindisch, V. Watson, "Robotically assisted nerve and facet blocks: A cadaver study", in *Journal of Academic Radiology*, vol. 9(7), pp. 821-825, 2002.

[33]   P. Corke, "Visual control of robot manipulators – a review", in *Visual Servoing*, Ed. K. Hashimoto, World Scientific, Singapore, vol. 7, pp. 1-31, 1993.

[34]   P. I. Corke, M. C. Good, "Dynamic effects in visual closed-loop systems", *IEEE Trans. Robotics and Automation*, vol. 5(12), 1996.

[35]   P. I. Corke, *Visual control of robots: high-performance visual servoing*, Research Studies Press Ltd., Somerset, England, 1996.

[36]   G. Corral , L. Ibanez, C. Nguyen, D. Stoianovici, Nassir Navab, K. Cleary, " Robot control by fluoroscopic guidance for minimally invasive spine procedures", to be published in Proc. *Computer Aided Radiology, CARS '04*, Chicago, 2004.

[37]   E. R. Cosman, "Development and technical features of the Cosman-Roberts-Wells (CRW) stereotactic system", in *Handbook of Stereotaxy*, Williams & Wilkins Publishers, Baltimore, USA, pp. 1-52, 1994.

[38]   B. Daly, P. A. Templeton, "Real-time CT fluoroscopy: evaluation of an interventional tool", in *Radiology*, vol. 211(2), pp. 309-315, 1999.

[39]   P. Dario, E. Guglielmelli, B. Allotta, M. C. Carrozza, "Robotics for medical applications", in *IEEE Robotics and Automation Magazine*, vol. 3(3), pp. 44-56, 1996.

[40]   B. L. Davies, R. D. Hibberd, M. J. Coptcoat, J. E. A. Wickham, "A surgeon robot for prostatectomy - a laboratory evaluation", in *J. Medical Engineering Technology*, vol. 13(6), pp. 273-277, 1989.

[41]   B. Davies, "From the guest editor", in *IEEE Engineering in Medicine and Biology*, vol. 14(3), p. 253, 1995.

[42]   B. L. Davies, "A discussion of safety issues for medical robots", in *Computer-Integrated Surgery - Technology and Clinical Applications*, R. H. Taylor et al. (Ed.), MIT Press, pp. 287-296, 1996.

[43]   B. Davis, "Safety of Robots in Surgery", in Proc. *2nd Workshop on Medical Robotics, IARP*, R. Dillmann, et al. (Ed.), p. 101, 1998.

[44]   B. Davies, "A review of robotics in surgery", in *Proc Inst Mech Eng*, vol. 214(1), pp. 129-140, 2000.

[45]   J. M. Drake, M. Joy, A. Goldenberg, D. Kreindler, "Computer and robotic assisted resection of brain tumors", in Proc. *5th International Conference on Advanced Robotics*, pp. 888-892, 1991.

[46]   J. F. Engelberger, *Robotics in practice*, IFS Publications Ltd., Kempston, England, 1980.

[47]   E. Euler, S. Wirth, K. J. Pfeifer, W. Mutschler, A. Hebecker, "3D-imaging with an isocentric mobile C-arm", in *electromedica*, vol. 68(2), pp. 122-126, 2000.

[48]   H. Fankhauser, D. Glauser, P. Flury, Y. Piguet, M. Epitaux, J. Favre, R. A. Meuli, "Robot for CT-guided stereotactic neurosurgery", in *Stereotactic Functional Neurosurgery*, vol. 63, pp. 93-98, 1994.

[49]   M. M. Favareto, "Polar 6000 - A new general purpose robot particularly suited for spot-welding applications", in *Proc. of the 4th Int. Conference on Industrial Robot Technology*, pp. 67-77, 1978.

[50]   G. Fichtinger, T. L. DeWeese, A. Patriciu, A. Tanacs, D. Mazilu, J. H. Anderson, K. Masamune, R. H. Taylor, D. Stoianovici, "System for robotically assisted prostate biopsy and therapy with intraoperative CT guidance", in *Journal of Academic Radiology*, vol. 9(1), pp. 60-74, 2002.

[51]   P. A. Finlay, M. H. Ornstein, "Controlling the movement of a surgical laparoscope - EndoSista", in *IEEE Engineering in Medicine and Biology*, vol. 14(3), pp. 289-291, 1995.

[52]   C. Frahm, W. Kloess, H.-B. Gehl, et al., "First experiments with a new laser-guidance device for MR-and CT-guided punctures", in *European Radiology*, vol. 5, p. 315, 1994.

[53]   J. J. Froelich, C. Scherf, "C-arm supported CT-Fluoroscopy", in *electromedica*, vol. 67(2), pp. 29-36, 1999.

[54]   J. J. Froelich, N. Ishaque, B. Saar, J. Regn, E. M. Walthers, F. Mauermann, K. J. Klose, "Control of percutaneous biopsy with CT-fluorscopy", in *Fortschritte Röntgenstrahlen*, vol. 170(2), pp. 191-197, 1999.

[55]   A. Gangi, B. Kastler, J. M. Arhan, A. Klinkert, et al. "A compact laser beam guidance system for interventional CT", in *Journal of Computer Assisted Tomography*, vol. 18, pp. 326-328, 1994.

[56]   D. Glauser, F. Flury, P. Durr, H. Fankhauser, C. W. Burckhardt, et al., "Configuration of a robot dedicated to stereotactic surgery", in *Stereotactic Functional Neurosurgery*, vol. 54, pp. 468-470, 1990.

[57]   D. Glauser, H. Frankenhauser, M. Epitaux, J.-L. Hefti, A. Jaccottet, "Neurosurgical robot Minerva, first results and current developments", in Proc. *of 2nd Symposium on MRCAS* , 1995.

[58]   P. L. Gleason, R. Kikinis, D. Altobelli, W. Wells, et al., "Video registration virtual reality for nonlinkage stereotactic surgery", in *Stereotactic Functional Neurosurgery*, vol. 63, pp. 139-143, 1996.

[59]   N. D. Glossop, R. W. Hu, J. A. Randle, "Computer-aided pedicle screw placement using frameless stereotaxis", in J. Spine, vol. 21(17), pp. 2026-2034, 1996.

[60]   B. S. Graves, J. Tullio, M. Shi, J. H. Downs, "An integrated remote neurosurgical system", *in Proc. of First Joint Conference of CVRMed and MRCAS,* Grenoble, Franc, pp. 799-808, 1997.

[61]   R. Greene, "Transthoracic needle aspiration biopsy", in *Interventional Radiology*, Ed. C. A. Athanasoulis, R. C. Pfister, Saunders, pp. 587-634, 1982.

[62]   W. Grimson, G. Ettinger, S. White, et al., "An automatic registration method for frameless stereotaxy, image guided surgery, and enhanced reality visualization", in *IEEE Trans. on Medical Imaging*, vol. 15, pp. 129-140, 1996.

[63]   R. W. Günther, G. Adam, P. Keulers, K. C. Klose, D. Vorwerk, "CT-gesteuerte Punktionen", in *Interventionelle Radiologie*, R. W. Günther et al. (Ed.), Thieme, Stuttgart, Germany, pp. 605-632, 1996.

[64]   K. C. Gupta, *Mechanics and control of robots*, Mechanical Engineering Series, Springer-Verlag New York Inc., New York, USA, 1997.

[65]   J. R. Haaga, "Interventional CT-guided procedures", in *Computed Thomography and Magnetic Resonance Imaging of the Whole Body*, J. R. Haaga et al. (Ed.), vol. 2(3), Mosby-Year Book, St. Louis, Missouri, USA, 1994.

[66]   R. M. Haralick, L. G. Shapiro, "Survey: Image segmentation techniques", in *J. Computer Vision, Graphics, and Image Processing*, vol. 29, pp. 100-132, 1985.

[67]   W. R. Hendee, "Realizing the true potential of medical imaging", in Radiology, vol. 209(3), pp. 604-605, 1998.

[68]   S. Hesse, *Industrieroboterpraxis: automatisierte Handhabung in der Fertigung*, Vieweg Verlag, Braunschweig/Wiesbaden, Germany, 1998.

[69]   V. A. Horsley, R. H. Clarke, "The structure and functions of the cerebellum examined by a new method", in *The Brain*, vol. 31, pp. 45-124, 1908.

[70]   J. Hsieh, "Analysis of the temporal response of computed tomography fluoroscopy", in Medical Physics, vol. 24(5), pp. 665-675, 1997.

[71]   J. Hummel, M. Figl, C. Kollmann, H. Bergmann, W. Birkfellner, " Evaluation of a miniature electromagnetic position tracker", in J. *Medical Physics*, vol. 29(10), pp. 2205-2212, 2002.

[72]   T. Irie, M. Kajitani, H. Yoshioka, K. Matsueda, Y. Inaba, et al., "CT fluoroscopy for lung nodule biopsy: A new device for needle placement", in *J. of Vascular Intervent. Radiol.*, vol. 11(3), pp. 359-364, 2000.

[73]   T. Irie, M. Kajitani, Y. Itai, "CT fluoroscopy-guided intervention: marked reduction of scattered radiation dose to the physician's hand by use of a lead plate and an improved I-I device", in J. *Vascular and Interventional Radiology*, vol. 12(12), pp. 1417-1421, 2001.

[74]   H. Ishijima, A. Taketomi, J. Aoki, K. Endo, Y. Koyama, N. Oya, N. Sato, "Applications of a combined angiography-CT system", in *electromedica*, vol. 68(1), pp. 65-74, 2000.

[75]   H. Ishizaka, T. Katsuya, Y. Koyama, H. Ishijima, T. Moteki, et al. "CT-guided percutaneous intervention using a simple laser director device", in *Annual Journal of Radiology*, vol. 170(3), pp. 745-746, 1998.

[76]   A. L. Jacob, P. Messmer, B. Baumann, N. Suhm, W. Steinbrich, P. Regazzoni, "Interactive single-step frameless freehand navigation in iliosacral screw fixation", in *Computer Assisted Orthopedic Surgery (CAOS)*, L. P. Nolte, et al. (Ed.), Hogrefe & Huber Publishers, Kirkland, USA, 1999.

[77]   J. Jankowski, W. Chruscielewski, J. Olszewski, M. Cygan, " System for personal dosimetry in interventional radiology", in J. *Radiation Protection Dosimetry*, vol. 101(1-4), pp. 221-224, 2002.

[78]   F. A. Jolesz, "Image-guided procedures and the operating room of the future", in *Radiology*, vol. 204, 601-612, 1997.

[79]   W. A. Kalender, *Computed Tomography – Fundamentals, System Technology, Image Quality, Applications*, Publicis MCD Verlag, Munich, Germany, 2000.

[80]   K. Katada, H. Anno, G. Takeshita, Y. Ogura, S. Koga, Y. Ida, K. Nonomura, T. Kanno, A. Ohashi, et al.,

"Development of real-time CT-fluoroscopy", in *Nippon Acta Radiologica*, vol. 54, pp. 1172-1174, 1994.

[81]   K. Katada, R. Kato, H. Anno, Y. Orgura, et al., "Guidance with real-time CT fluoroscopy: Early clinical experience", in Radiology, vol. 200(3), pp. 851-856, 1996.

[82]   R. Kato, K. Katada, H. Anno, S. Suzuki, Y. Ida, S. Koga, "Radiation dosimetry at CT fluoroscopy: physicians hand dose and development of needle holders", in *Radiology*, vol. 201(2), pp. 576-578, 1996.

[83]   R. Klöppel, T. Friedrich, U. Eichfeld, W. Wilke, T. Kahn, "CT-guided marking of pulmonary lesions before thoracoscopic resection", in *J. Der Radiologe*, vol. 41, pp. 201-204, 2001.

[84]   M. S. Konstantinov, Z. I. Zankov, "Multi-grippers hot forge manipulators", in *Industrial Robots*, vol. 2, pp. 47-55, 1975.

[85]   E. Kreyszig, *Advanced Engineering Mathematics*, John Wiley & Sons, Inc., 1988.

[86]   On web-page of R. Kumar: *http://www.cs.jhu.edu/~rajesh/*

[87]   Y. S. Kwoh, I. S. Reed, J. Y. Chen, H. M. Shao, T. K. Truong, E. Jonckheere, "A new computerized tomographic-aided robotic stereotaxis system", in *Robotics Age*, vol. 7, pp. 17-22, 1985.

[88]   Y. S. Kwoh, J. Hou, E. A. Jonckheere, S. Hayati, "A robot with improved absolute positioning accuracy for CT guided stereotactic brain surgery", in *IEEE Transactions on Biomedical Engineering*, vol. 35(2), pp. 153-160, 1988.

[89]   K. Lackner, P. Landwehr, K.-H. Schlolaut, T. Feyerabend, "CT-gesteuerte Punktionen", in Angioplastie, Embolisation, Punktion, Drainagen - Interventionelle Methoden der Radiologie, G. Friedmann et al. (Ed.), Schnetztor-Verlag, Konstanz, Germany, pp. 149-161, 1989.

[90]   R. E. Latchaw, "Neuroradiology research: The opportunities and the challenges", in *Radiology*, vol. 209, pp. 3-7, 1998.

[91]   S. Lavallée, J. Troccaz, L. Gaborit, et al., "Image guided operating robot: a clinical application in stereotactic neurosurgery", in *Proc. IEEE Int. Conf. on Robotics & Automat.*, vol. 92, pp. 618-624, 1992.

[92]   S. Lavallée, R. Szeliski, "Recovering the position and orientation of free-form objects from image contours using 3D distance maps", in *IEEE Trans. PAMI*, vol. 17(4), pp. 378-390, 1995.

[93]   S. Lavallée, "Registration for computer-integrated surgery: methodology, state of the art", in *Computer-Integrated Surgery – Technology and Clinical Applications*, R. H. Taylor, et al. (Ed.), MIT Press, Cambridge, USA, pp. 77-97, 1996.

[94]   S. Lavallée, R. Szeliski, L. Brunie, "Anatomy-based registration of 3D medical images, range images, X-ray projections and 3D models using octree-splines", in *Computer-Integrated Surgery – Technology and Clinical Applications*, R. H. Taylor, et al. (Ed.), MIT Press, Cambridge, USA, pp. 115-143, 1996.

[95]   C. Lee, Y. F. Wang, D. R. Uecker, Y. Wang, "Image analysis for automated tracking in robot-assisted endoscopic surgery", in Proc. *Int. Conf. on Pattern Recognition*, pp. 88-92, 1994.

[96]   D. Loulmet, A. Carpentier, N. d'Attellis, et al., "Endoscopic coronary artery bypass grafting with the aid of robotic assisted instruments", in *J. Thoracic and Cardiovascular Surgery*, vol. 118(1), pp. 4-10, 1999.

[97]   T. C. Lueth, E. Heissler, et al., "Robotics for hyperthermia and surgical applications", in *Proc. of 2nd Workshop Med. Robotics, IARP*, Heidelberg, Germany, R. Dillmann et al. (Ed.), pp. 80-89, 1997.

[98]   T. C. Lueth, A. Hein, J. Albrecht, M. Demirtas, S. Zachow, et al., "A surgical robot system for maxillo-facial surgery", in *IEEE Int. Conference on Industrial Electronics (IECON)*, pp. 2470-2475, 1998.

[99]   R. J. Maciunas, "Overview of interactive image-guided neurosurgery: principles, applications, and new techniques", in *Advanced Neurosurgical Navigation*, E. Alexander III, et al. (Ed.), Thieme Medical Publishers, Inc., New York, pp. 15-32, 1999.

[100]   A. R. Margulis, J. H. Sunshine, "Radiology at the turn of the Millenium", in *Radiology*, vol. 214, pp. 15-23, 2000.

[101]   K. Masamune, et al., "A newly developed stereotactic robot with detachable driver for neurosurgery", in *Proceedings Medical Image Computing and Computer Ass. Interv., MICCAI 1998*, pp. 215-222, 1998.

[102]   K. Masamune, G. Fichtinger, A. Patriciu, R. C. Susil, R. H. Taylor, L. R. Kavoussi, J. H. Anderson, I. Sakuma, T. Dohi, S. Stoianovici, "System for robotically assisted percutaneous procedures with computed tomography guidance", in *Journal Computer Aided Surgery*, vol. 6(6), pp. 370-383, 2001.

[103]   A. McCarthy, "Robotics hearing with spoken commands", Stanford Artificial Intelligence Laboratory, Stanford University, AIM 56, 1967.

[104]   A. Melzer, A. Schmidt, K. Kipfmüller, D. Grönemeyer, R. Seibel, "Technology and principles of tomographic image-guided interventions and surgery", in *Surg Endosc*, vol. 11, pp. 946-956, 1997.

[105]   L. Mettler, M. Ibrahim, W. Jonat, "One year of experience working with the aid of a robotic assitant (the voice-controlled optic holder AESOP)", in *Human Reproduction*, vol. 13(10), pp. 2748-2750, 1998.

[106]   B. D. Mittelstadt, P. Kazanzides, J. F. Zuhars, B. Williamson, P. Cain, et al., "The evolution of a surgical robot from prototype to human clinical use", in *Computer-Integrated Surgery - Technologie and Clinical Applications*, R. H. Taylor et al. (Ed.), The MIT Press, Cambridge, MA, USA, pp. 257-262, 1996.

[107]   E. N. Moore, *Theoretical Mathematics*, John Wiley & Sons, Inc., 1983.

[108]   M. Mori, N. Nitta, K. Murata, T. Sakamoto, R. Morita, "A newly developed equipment for biopsy under CT fluoroscopy", in *American Journal of Roentgenology*, vol. 170, p. 128, 1998.

[109]   S. K. Mukherji, "Head and neck imaging: The next 10 years", in Radiology, vol. 209, pp. 8-14, 1998.

171

[110] N. Navab, B. Geiger, "A simple intra-operative imaging guided positioning device for needle biopsy", Technical Report: Siemens Corporate Research, 755 College Road East, Princeton, NJ, USA, 1996.

[111] R. Nevatia, "Depth measurement by motion stereo", *Computer Graph. and Image Process.*, vol. 5, pp. 203-214, 1976.

[112] J. H. Newhouse, R. C. Pfister, "Renal cyst puncture", in *Interventional Radiology*, Ed. C. A. Athanasoulis, R. C. Pfister, Saunders, pp. 409-425, 1982.

[113] N. Nitta, M. Mori, K. Murata, M. Takahashi, A. Mishina, et al., "A new mechanically manipulated unit for CT-guided biopsy", in *Nippon Acta Radiologica*, vol. 57(11), pp. 675-677, 1997.

[114] L. P. Nolte, R. Ganz, *Computer Assisted Orthopedic Surgery (CAOS)*, Hogrefe & Huber Publishers, Kirkland, USA, 1999.

[115] G. Onik, P. Costello, E. Cosman, et al., "CT body stereotaxis: an aid for CT-guided biopsies", in *Annual Journal of Radiology*, vol. 146, pp. 163-168, 1986.

[116] R. C. Otto, R. F. Dondelinger, J. C. Kurdziel, "Abdominal Biopsy", in *Interventional Radiology*, R. F. Dondelinger et al. (Ed.), Thieme, Stuttgart, Germany, pp. 33-52, 1990.

[117] A. M. Palestrant, "Comprehensive approach to CT-guided procedures with a hand-held guidance device", in *Radiology*, vol. 174, pp. 270-272, 1990.

[118] J. R. Parker, *Algorithms for Image Processing and Computer Vision*, Wiley Computer Publishers, New York, 1997.

[119] A. Patriciu, D. Stoianovici, L. L. Whitcomb, T. Jarrett, D. Mazilu, A. Stanimir, I. Iordachita, J. Anderson, R. H. Taylor, et al., "Motion-based robotic instrument targeting under C-arm fluoroscopy", in Proc. *Medical Image Computing and Computer Assisted Intervention, MICCAI'00*, pp. 988-996, Oct. 2000.

[120] A. Patriciu, S. Solomon, L. Kavoussi, D. Stoianovici, "Robotic kidney and spine percutaneous procedures using a new laser-based CT registration method", in Proc. *Medical Image Computing and Computer Assisted Intervention, MICCAI'01*, pp. 249-257, Oct. 2001.

[121] A. Patriciu, D. Mazilu, D. Petrisor, L. Kavoussi, D. Stoianovici, "Automatic targeting method and accuracy study in robot assisted needle procedures", in Proc. *Medical Image Computing and Computer Assisted Intervention*, MICCAI'03, pp. 124-131, 2003.

[122] A. Paul, " Surgical robot in endoprosthetics - How CASPAR assists on the hip", in *MMW-Fortschr. Med.*, vol. 141(33), p. 18, 1999.

[123] *PinPoint™ - CT Scanner Integrated Stereotactic Arm*, Product Data, Picker International, Inc., Cleveland, USA, 1998.

[124] I. Pitas, *Digital Image Processing Algorithms and Applications*, Wiley, New York, 2000.

[125] P. Potamianos, B. L. Davies, R. D. Hibberd, "Intra-operative image guidance for keyhole surgery: methodology and calibration", in Proc. *1st Int. Symposium on Medical Robotics and Computer Assisted Surgery, MRCAS*, Pittsburgh, pp. 98-104, 1994.

[126] P. Potamianos, B. L. Davies, R. D. Hibberd, "Intra-operative registration for percutaneous surgery", in Proc. *2nd Int. Symposium on Medical Robotics and Computer Assisted Surgery, MRCAS*, Baltimore, pp. 156-164, 1995.

[127] B. K. Poulose, M. F. Kutka, M. Mendoza-Sagaon, A. C. Barnes, C. Yang, R. H. Taylor, M. A. Talamini, "Human versus robotic organ retraction during laparoscopic Nissen fundoplication", in *Surgical Endoscopy*, vol. 13(5), pp. 461-465, 1999.

[128] G. Rau, K. Radermacher, B. Thull, C. von Pichler, "Aspects of ergonomic system design applied to medical work systems", in *Computer-Integrated Surgery – Technology and Clinical Applications*, R. H. Taylor, et al. (Ed.), MIT Press, Cambridge, USA, pp. 203-221, 1996.

[129] H. Reichenspurner, R. Damiano, M. Mack, D. Boehm, H. Gulbins, et al., "Use of the voice-controlled and computer-assisted surgical system ZEUS for endoscopic coronary artery bypass grafting", in *J. Thoracic and Cardiovascular Surgery*, vol. 118(1), pp. 11-16, 1999.

[130] J. M. Sackier, Y. Wang, "Robotically assisted laparoscopic surgery - From concept to development", in *Journal of Surgical Endoscopy*, vol. 8, pp. 63-66, 1994.

[131] S. E. Salcudean, G. Bell, S. Bachmann, W. H. Zhu, P. Abolmaesumi, P. D. Lawrence, "Robot-assisted diagnostic ultrasound – design and feasibility experiments", Conf. on Medical Image Computation and Computer-Assisted Intervetion, pp. 1062-1071, 1999.

[132] J. J. Santos-Munné, M. A. Peshkin, S. Mirkovic, S. D. Stulberg, T. C. Kienzle, "A stereotactic and robotic system for pedicle screw placement", in Proc. *Medicine Meets Virtual Reality III Conference*, San Diego, pp. 326-333, 1995.

[133] S. J. Savader. C. A. Prescott, G. B. Lund, F. A. Osterman, "Inraductal bilary biopsy: Comparison of three techniques", in *Journal of Vascular and Interventional Radiology*, vol. 7(5), pp. 743-750, 1996.

[134] A. M. Schmidt, A. Melzer, R. M. Seibel, D. H. Grönemeyer, "Tomographic guided biopsy and cell aspiration of neoplasms", in *Minimally Invasive Therapy & Allied Technologies*, vol. 5, pp. 249-254, 1996.

[135] S. Schreiner, J. Funda, A. C. Barnes, J. H. Anderson, "Accuracy assessment of a clinical biplane fluoroscope for 3D measurements and targeting", in Proc. *SPIE Medical Imaging Conference 1997: Image Display*, vol. 3031, pp. 160-166, 1997.

172   References

[136] B. Schueler, X. Hu, "Correction of image intensifier distortion for three-dimensional X-ray angiography", in Proc. *SPIE Medical Imaging Conference*, vol. 2432, pp. 272-279, 1995.

[137] M. O. Schurr, A. Arezzo, B. Neisius, H. Rininsland, H. U. Hilzinger, et al., "Trocar and instrument positioning system TISKA", in *Surgical Endoscopy*, vol. 13(5), pp. 528-531, 1999.

[138] G. D. Schweiger, "Computed tomography fluoroscopy: Techniques and applications", in *Current Problems in Diagnostic Radiology*, vol. 29(1), pp. 1-26, 2000.

[139] G. D. Schweiger, B. P. Brown, R. E. Pelsang, et al., "CT fluoroscopy: technique and utility in guiding biopsies of transiently enhancing hepatic masses", in *Abdominal Imaging*, vol. 25(1), pp. 81-85, 2000.

[140] R. M. M. Seibel, D. H. W. Grönemeyer, *Interventional computed tomography*, Blackwell Scientific Publications, Cambridge, MA, USA, 1990.

[141] R. M. Seibel, C. Sehnert, J. Plassmann, A. M. Schmidt, "Interventional procedures with real time CT", in *Radiology*, vol. 205, p. 383, 1997.

[142] M. Shahinpoor, *A robot engineering textbook*, Harper & Row Publishers Inc., New York, USA, 1987.

[143] I. H. Shames, *Engineering mechanics - Static and dynamics*, Prentice-Hall, Inc., 1997.

[144] D. H. Sheafor, E. K. Paulson, M. A. Kliewer, D. M. DeLong, R. C. Nelson, "Comparison of sonographic and CT guidance techniques: does CT fluoroscopy decrease procedure time ?", in *American J. of Radiology*, vol. 174(2), pp. 939-942, 2000.

[145] H. Shennib, A. Bastawisy, J. McLoughlin, et al., "Robotic computer-assisted telemanipulation enhances coronary artery bypass", in *J. Thoracic and Cardiovascular Surgery*, vol. 117(2), pp. 310-313, 1998.

[146] S. G. Silverman, K. Tuncali, D. F. Adams, R. D. Newfel, et al., "CT fluoroscopy-guided abdominal interventions: Techniques, results, and radiation exposure", in *Radiology*, vol. 212(3), pp. 673-681, 1999.

[147] D. A. Simon, M. Herbert, T. Kanade, "Techniques for fast and accurate intra-surgical registration", in *J. of Image Guided Surgery*, vol. 1(1), pp. 17-29, 1995.

[148] W. N. Sinner, *Needle biopsy and transbronchial biopsy*, Thieme, Stuttgart, Germany, pp. 35-53, 1982.

[149] N. Sohn, R. D. Robbins, "Computer-assisted surgery", in *New England J. of Medicine*, vol. 312(14), p. 924, 1985.

[150] S. B. Solomon, P. White, C. M. Wiener, J. B. Orens, K. P. Wang, "3D CT-guided bronchoscopy with a real-time electromagnetic position sensor", in J. *Chest*, vol. 118(6), pp. 1783-1787, 2000.

[151] A. H. Sonin, B. Penrod, E. Owens-Brown, "CT-fluorscopic guidance of sacroiliac joint injection", in *American J. of Roentgenology*, vol. 170, p. 38, 1998.

[152] D. Stoianovici, J. A. Cadeddu, R. D. Demaree, S. A. Basile, R.H. Taylor, et al., "A novel mechanical transmission applied to percutaneous renal access", in Proc. *ASME Dynamic Systems and Control Division, DSC*, vol. 61, pp. 401-406, 1997.

[153] D. Stoianovici, J. A. Cadeddu, R. D. Demaree, S. A. Basile, R. H. Taylor, L. L. Whitcomb, et al., "An efficient needle injection technique and radiological guidance method for percutaneous procedures", in Proc. *1st Joint Conf. of CVRMed and MRCAS,* Grenoble, France, pp. 295-298, 1997.

[154] D. Stoianovici, L. L. Whitcomb, J. H. Anderson, R. H. Taylor, L. R. Kavoussi, "A modular surgical robotic system for image guided percutaneous procedures", in Proc. *Medical Image Computation and Computer-Assisted Intervention*, pp. 404-410, 1998.

[155] L. M. Su, D. Stoianovici, T. W. Jarrett, A. Patriciu, W. W. Roberts, J. A. Dadeddu, S. Ramakumar, et al., "Robotic percutaneous access to the kidney: comparison with standard manual access", in Journal of Endourology, vol. 16(7), pp. 471-475, 2002.

[156] R. C. Susil, J. H. Anderson, R. H. Taylor, "A single image registration method for CT guided interventions", in *Proceedings MICCAI 1999*, pp. 798-808 , 1999.

[157] R. H. Tylor, B. D. Mittelstadt, H. A. Paul, W. Hanson, et al., "An image-directed robotic system for precise orthopedic surgery", in *IEEE Trans. on Robotics and Automation*, vol. 10(3), pp. 261-275, 1994.

[158] R. H. Taylor, J. Funda, B. Eldridge, S. Gomory, K. Gruben, D. LaRose, et al., "A telerobotic assistant for laparoscopic surgery", in *IEEE Engineering in Medicine and Biology*, vol. 14(3), pp. 279-288, 1995.

[159] R. H. Taylor, J. Funda, L. Joskowicz, A. D. Kalvin, S. H. Gomory, A. P. Gueziec, L. M. G. Brown, "An overview of computer-integrated surgery at the IBM Thomas J. Watson research center", in *IBM Journal of Research and Development*, vol. 40(2), pp. 163-183, 1996.

[160] R. H. Taylor, P. Jensen, L. Whitcomb, A. Barnes, R. Kumar, D. Stoianovici, P. Gupta, Z. X. Wang, E. deJuan, L. Kavoussi, "A steady-hand robotic system for microsurgical augmentation", in *J. Robotics Research*, vol. 18(12), pp. 1201-1210, 1999.

[161] V. Urban, M. Wapler, T. Weisener, R. Schönmayr, "A tactile feedback hexapod operating robot for endoscopic procedures", in *Neurological Research*, vol. 21, pp. 28-30, 1999.

[162] U. Voges, E. Holler, B. Neisius, M. Schurr, T. Vollmer, "Evaluation of ARTEMIS, the advanved robotics and telemanipulator system for minimally invasive surgery", in Proc. *2nd Workshop on Medical Robotics, IARP*, Heidelberg, Germany, R. Dillmann et al. (Ed.), pp. 137-148, 1997.

[163] J. P. Wadley, N. L. Dorward, M. Breeuwer, F. A. Geritsen, N. D. Kitchen, D. G. T. Thomas, "Neuronavigation in 210 cases: further development of applications and full integration into contemporary neu-

rosurgical practice", in *CAR '98*, H. U. Lemke et al. (Ed.), Elsevier Science, pp. 635-640, 1998.

[164] H.-J. Wagner, H. Theis, R. Zitzmann, "Endovascular therapy unit: Angiography and intervention systems under OR conditions", in *electromedica*, vol. 67(2), pp. 23-28, 1999.

[165] G. Wang, G. Schweiger, M. W. Vannier, "An interative algorithm for X-ray CT-fluoroscopy", in IEEE Transactions on Medical Imaging, vol. 17(5), pp. 853-856, 1998.

[166] J. Wang, W. J. Wilson, "3D relative position and orientation estimation using Kalman filtering for robot control", Wiley, New York, pp. 2638-2645, 1992.

[167] M. Wapler, J. Stallkamp, et al., "Motion feedback as a navigation aid in robot assisted neurosurgery", in *Medicine Meets Virtual Reality*, J. D. Westwood, et al. (Ed.), IOS Press, pp. 215-219, 1998.

[168] E. Watanabe, T. Watanabe, S. Manaka, et al., "Three-dimensional digitizer (neuronavigator): new equipment for CT-guided stereotactic surgery", in *Surgical Neurology*, vol. 27, pp. 543-547, 1987.

[169] E. Watanabe, "The Neuronavigator: A computer-controlled navigation system in neurosurgery", in *Computer-Integrated Surgery - Technology and Clinical Applications*, R. H. Taylor, et al. (Ed.), MIT Press, Cambridge, USA, pp. 319-327, 1996.

[170] G.-Q. Wei, K. Arbter, G. Hirzinger, "Real-time visual servoing for laparoscopic surgery", IEEE Journal *Engineering in Medicine and Biology*, vol. 16(1), pp. 40-45, 1997.

[171] L. E. Weiss, A. C. Sanderson, C. P. Neumann, "Dynamic sensor-based control of robots with visual feedback", IEEE Journal *Robotics and Automation*, vol. RA-3, pp. 404-417, 1987.

[172] J. L. Westcott, "Lung biopsy", in *Interventional Radiology*, Ed. R. F. Dondelinger et al, Thieme Medical Publishers, Inc., New York, pp. 9-17, 1990.

[173] D. B. Westmore, W. J. Wilson, "Direct dynamic control of a robot using an end-point mounted camera and Kalman filter position estimation", in Proc. *IEEE Int. Conf. on Robotics and Automation*, pp. 2376-2384, 1991.

[174] E. J. Wiesen, F. Miraldi, "Imaging Principles in Computed Tomography", in *Computed Tomography and Magnetic Resonance Imaging of The Whole Body*, J. R. Haaga et al. (Ed.), vol. 1(3), Mosby-Year Book, St. Louis, Missouri, USA, 1994.

[175] B. Williamson, Jr., "The electronic transformation of radiology", in *Radiology*, vol. 209(3), pp. 606-608, 1998.

[176] D. J. Wilson, "Abscess drainage", in *Interventional Radiology of the Musculoskeletal System*, D. J. Wilson (Ed.), Edward Arnold, London, UK, pp. 21-32, 1995.

[177] P. R. Wolf, *Elements of Photogrammetry*, McGraw-Hill, 1974.

[178] Y. Yakimovsky, R. Cunningham, "A system for extracting three-dimensional measurements from a stereo pair of TV cameras", in *Computer Graphics and Image Processing*, vol. 7, pp. 195-210, 1978.

[179] J. Yao, R. H. Taylor, R. P. Goldberg, R. Kumar, A. Bzostek, R. Van Vorhis, et al., "A progressive cut refinement scheme for revision total hip replacement surgery using C-arm fluoroscopy", in Proc. *Medical Image Computation and Computer-Assisted Intervention, MICCAI'99*, pp. 1010-1019, 1999.

[180] R. F. Young, "Application of robotics to stereotactic neurosurgery", in *Neurosurgical Research*, vol. 9, pp. 123-128, 1987.

[181] L. Zamorano, A. Pandya, Q. H. Li, R. Pérez-de la Torre, P. Pittet, F. Badano, V. Robert, "The clinical use and accuracy of the Neuromate robot for open neurosurgery", in Proc. *Computer Assisted Radiology and Surgery, CAR '98*, H. U. Lemke et al. (Ed.), pp. 185-190, 2000.

[182] *Medical imaging - the assessment of image quality*, International Commission on Radiation Units and Measurements, Inc. (ICRU), report 54, 1995.

[183] C-842 DC-Motor Controller, Operating Manual, Polytec PI, Inc., Tustin, CA, USA, 1999.

[184] "Safety", in *Computer-Integrated Surgery – Technology and Clinical Applications*, R. H. Taylor et al. (Ed.), MIT Press, pp. 283-285, 1996.

Appendix

Functions for hardware initialization:	
InitSiemon	initialization of motor and encoder PC-cards
ExitSiemon	close the hardware communication and deallocates the memory

Functions for robot motion:	
MoveSiemonA	move needle holder absolute in Euler-coordinates
MoveSiemonR	move needle holder relative in Euler-coordinates
MoveSiemonR_without_waiting	see above, but function gives control back before desired position is reached
MoveSiemonMotorsA	move motors absolute (direct control of both robot main axes)
MoveSiemonMotorsR	move motors relative (direct control of both robot main axes)
MoveSiemonToCounter	move specified motor to its desired axis-encoder position
KeepPosition	needle is actively kept in the desired position (active backlash compensation)
CorrectPosition	correction of an eventual deviation between motor- and axis-encoder values
FindLimitsAndGoHome	motor moves till limit switch is reached, then moves back to home position
DefineHome	current position is defined as home position
GoHome	move specified motor to home position
SetSpeedOfSiemon	define the maximum speed of the specified motor
StopSiemon	sudden stop of all motors

Functions for needle control:	
NeedleActiv	activation/deactivation of the needle drive
MoveNeedleA	move needle absolute for needle insertion
MoveNeedleA_without_waiting	see above, but function gives control back before desired position is reached
MoveNeedleR	move needle relative for needle insertion
MoveActiveNeedleR	rotates the inner and outer cannula of the "active needle"

Functions for position and encoder signal capturing:	
GetAbsoluteAngles	get the desired absolute Euler-angles of the needle cannula
GetAxisEncoderAngles	get both the current axes-encoder angles (robot axis-encoder)
GetMotorEncoderAngles	get both the current motor-encoder angles (motor-encoder)
GetCounter	get count of specified axis-encoder
GetEulerAnglesFromAxisEncoders	compute Euler-angles out of robot axes-encoder values
GetEulerAnglesFromMotorEncoders	compute Euler-angles out of motor-encoder values
GetEncoderNeedleVector	compute needle direction vector out of robot axes-encoder values

Functions for table-simulator motion (linear drive):	
MoveTableSimulatorR	move table simulator (linear drive) relative
ZeroTableSimulatorPos	define current table simulator position as home (zero) position

Functions for CT-scanner control:	
CTImaging	activation/deactivation of radiation (equivalent to pressing the CT foot pedal)
MoveCTTableR	move CT-table incrementally with a specified number of steps

Functions for C-arm control:	
InitializeFluoro	initialization of the C-arm for radiation
FluoroImaging	activation/deactivation of radiation (equivalent to pressing the C-arm foot pedal)

Table 12.1: Function reference "SiemonClass" (robot control)

Functions for hardware initialization:	
InitFramegrabber	initialization of frame grabber PC-card
ExitFramegrabber	close the hardware communication and deallocate the image memory
Functions for image acquisition:	
GetImage	copies the current video image (CT image, input 1) into the image memory
GetImage (2)	see above, additionally the target point is given to the function/assigned
GetImage (3)	see above, additionally the min. and max. grayscale in the image is returned
GetCameraImage	copies the current video image (Video camera, input 2) into the image memory
SetBrightnessContrast	sets the brightness and contrast parameters for the frame grabber
WaitForFirstCTImage	function is waiting till the first CT-image is provided at video input and returns image
Functiones for image analysis and image feature detection:	
DetectNeedle	detects needle and needle parameters in whole image (location, orientation, etc.)
DetectNeedleROI	detects needle (location, orientation, etc.) in specified "Region Of Interest" (ROI)
FluoroDetectNeedle	detects needle (location, orientation, etc.) in X-ray fluoroscopy image
NumOfPixel	returns the number of pixels assigned to the needle in a specified ROI
CatchPoint	returns the "center of gravity" of a contiguous cluster of pixels (threshold based)
CatchPoint (2)	returns the "center of gravity" of all pixels with grayscale > threshold in ROI
Further image functions:	
DrawImage	displays image on the screen
SaveImage	saves image data on hard disc
LineToBorder	draws a line on the displayed image (line origin, direction vector)
DrawTargetMarker	draws a crosshair marker on the displayed image
DeleteTargetMarker	deletes a crosshair marker

Table 12.2: Function reference "ImageClass" (frame grabber control, image acquisition)

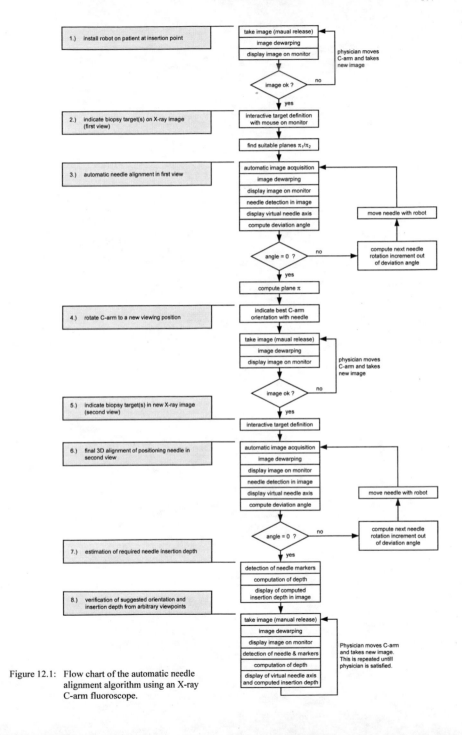

Figure 12.1: Flow chart of the automatic needle
alignment algorithm using an X-ray
C-arm fluoroscope.